J. Solberg

ERIC HOMBERGER

THE PENGUIN
HISTORICAL ATLAS OF NORTH AMERICA

PENGUIN BOOKS

Published by the Penguin Group
Penguin Books Ltd, 27 Wrights Lane, London W8 5TZ, England
Penguin Books USA Inc., 375 Hudson Street, New York, NY 10014, USA
Penguin Books Australia Ltd, Ringwood, Victoria, Australia
Penguin Books Canada Ltd, 10 Alcorn Avenue, Toronto, Ontario, Canada M4V 3B2
Penguin Books (NZ) Ltd, 182–190 Wairau Road, Auckland 10, New Zealand

Penguin Books Ltd, Registered Offices: Harmondsworth, Middlesex, England

First published 1995
Published simultaneously by Viking
1 3 5 7 9 10 8 6 4 2

Printed and bound in Great Britain by The Bath Press, Avon

ISBN 0–14–0–51327–2

Foreword

In the case of the United States, Mexico and Canada, nations newly arrived upon the historical stage, the story of the making of the nation has been the overwhelming concern of historians. Across the stage strut politicians and generals, robber barons and high-minded patriots, while in the remote background one glimpses the masses, the anonymous footsoldiers of history. The "new" American history—not so new anymore—sought to open the subject to the experience of groups generally excluded by the historians' preoccupation with political, diplomatic and military affairs. Atlases are among the most conservative of historical forms, and remain, if one may personify that diverse category, quite insistent upon the old concerns. I have been intrigued by the general difficulties of mapping certain kinds of historical experience. There are detailed maps of the battle of Midway, and of Pickett's charge at Gettysburg, but not of the changing configuration of domestic space.

The notion of a regional historical atlas thus carries with it another kind of challenge. It invites us to think of the interdependency of peoples, of the tensions and also the peaceful resolutions and accommodations which enable diverse traditions to endure. But warfare, and not peaceful co-existence, forms a substantial theme running through this atlas. The St. Lawrence River, the Great Lakes and the Rio Grande have been traditional battlegrounds. The extensive economic interdependence, recognised by the free trade agreement of 1993, makes ever more sensible the willingness to think of the land from Alaska to the Yucatan Peninsula as constituting a whole.

My debts to scholarship are deep and warmly acknowledged. Richard B. Morris, Encylopedia of American History (1953), Michael C. Meyer and William L. Sherman, The Course of Mexican History (1979; 4th ed. 1991), and J.M. Bumsted's magisterial The Peoples of Canada: A Pre-Confederation History (1992) have been my basic tools. I have shamelessly drawn upon the expertise of Roger Thompson and Jacqueline Fear-Segal, colleagues at the University of East Anglia. My editor, Caroline Lucas, has shown saintly fortitude at the vagaries of her author.

Eric Homberger
Norwich, April 1995

This book is dedicated to my son, Martin Homberger.
I hope someday we will get to do that book together.

Contents

Timeline: 40,000 BC to AD 1179

CANADA

40–30,000 possible first arrivals from Asia across Bering land bridge

13,000 Bluefish Cave occupied—earlier occupation site in the Yukon and Alaska region

MEXICO

21,000 evidence of occupation of Tlapacoya, near Mexico City

11,500 beginning of Olmec civilization in Mexico

5,000 first cultivation of maize in Tehuacán Valley, Mexico

2,500 ceramics in use in Mesoamerica

1,500 Highland Olmec settlement at Chalcatzingo
1,200 beginning of Formative period in Mesoamerica, marked by the first urban civilizations, particularly of the Olmec

900 La Venta replaces San Lorenzo as the principal centre of the Mesoamerican Olmec

500–400 ceremonial demolition of La Venta
300 rise of the Maya civilization in Mesoamerica
200 rise of Teotihuacán

AD 50 City of Teotihuacán, central Mexico, laid out on a regular grid plan. Construction of Pyramid of the Sun—largest structure in the Americas before advent of the Europeans

300 Classic period of Maya civilization begins

500 Teotihuacán is the world's sixth largest city, with a population of c. 200,000
600 apogee of Maya civilization
650 sudden abandonment of Teotihuacán

800 development of Toltec urban centre at Tula
850 collapse of classic Maya culture in Mesoamerica

950 Toltecs rise to power in Central Mexico

1150 collapse of Tula
1156–88 invasion and conquest of Yucatán by Toltec survivors from the fall of Tula
1179 fall of Toltecs in central Mexico

NORTH AMERICA AND U.S.A.

12,000 first secure traces of human occupation in North America, in Meadowcroft Rock, Pennsylvania

10,000 North American ice-sheets retreat and settlements become more abundant. Palaeo-Indian stage in North America. Nomadic big-game hunters using first Clovis and later Folsom stone projectile points

8,000 Archaic stage in North America. Increased variety of food sources—smaller game, plants, fish and shellfish. Development of new tools

7,000 semi-permanent settlements emerge in North America. Increasing use of plant foods in diet

5,000 use of bottle gourd as a container commences in eastern North America. Mesoamerican in origin

2,000 most hunter-gatherer tribes engage in agriculture

1,500 Woodlands, Ohio–Mississippian culture. Settled villages

1,000 beginning of the Adena culture in the eastern woodlands of North America, characterized by rich mound burials

300 rise of Hopewell Indian chiefdoms in North America

AD 100 Mogollon/Hohokam culture in Southwest

200–900 occupation of Monte Albán by Zapotecs

500 Basketmaker III period in American Southwest
600 Mississippian temple mounds and tombs

800 first use of bow and arrow in Mississippi Valley

900 Hohokam build irrigation channels up to 16 km. Influence from Mexico revealed in imports and building style

1000 Vikings colonize Greenland and discover America, founding short-lived settlements on coasts of Labrador and Newfoundland

1050 Anasazi settlements of the North American Southwest move to well-defended positions, e.g. Pueblo Bonita in Chaco Canyon
1100 Thule culture of Arctic America, with whaling-based economy, expands eastwards, dominating area from Greenland to Siberia

Timeline: 1200 to 1664

CANADA

1534–35 Jacques Cartier explores St. Lawrence River

1542 Cartier's colony at Québec abandoned

1550s French fur traders penetrate Gulf of St. Lawrence

1599 first French dwelling in Canada built at Tadoussac

1608 Québec founded by the French

1642 Montréal founded as a mission

1649 Iroquois league conquers the Hurons

1663 Québec reorganized as a crown colony of France

MEXICO

1200 Aztecs occupy Valley of Mexico

1345 Aztecs colonize central Mexico, founding capital, Tenochtitlán

1428 Aztec Empire formed
1440–68 reign of Aztec ruler, Montezuma I

1519 Spaniards under Hernán Cortés arrive in central Mexico. Aztecs overthrown in two years
1524 twelve Franciscan friars arrive in New Spain

1539 first book, a religious tract in Náhuatl and Spanish, published in New Spain
1540 Coronado explores the Southwest
1542 De Soto explores the Southwest
1545 silver discovered at Zacatecas, Mexico
1546 completion of the Spanish conquest of the Maya
1549 Spanish crown abolishes *encomienda* in Mexico

1563 work begins on the cathedral in Mexico City
1565 Spanish found first permanent settlement north of Mexico at St. Augustine, Florida
1574 inquisition in Mexico: first *Auto da Fé*

1593 Alameda Park laid out in Mexico City

NORTH AMERICA AND U.S.A.

1350 collapse of Pueblo cultures due to warfare or climatic decline

1450 depopulation and abandonment of towns in the Middle Mississippi area—possibly due to disease

1492 Columbus discovers the New World
1493 Hispaniola founded—the first Spanish settlement in the New World. Treaty of Tordesillas divides New World between Spain and Portugal
1497 John Cabot reaches North America

1510 first African slaves transported to the Americas
1513 Ponce de Léon establishes Spanish claim to Florida

1524 Giovanni Verrazano explores coast of North America

1585 first British colonies established at Roanoke Island, North Carolina

1607 first English settlement in the Americas: at Jamestown, Virginia
1609 Henry Hudson claims part of North America for United Provinces. "Starvation winter" in Jamestown
1612 John Rolfe plants first tobacco in Virginia
1619 first Negroes brought to British America. Virginia begins representative assembly
1620 Plymouth colony founded by English Puritans

1625 Dutch found first permanent settlement on Manhattan

1635 founding of Connecticut colony
1636 Roger Williams founds Rhode Island colony. Harvard College founded
1638 Pequot war in New England. Swedish settlement founded on Delaware River

1660s legal definition of negro slavery first used in Virginia and Maryland
1663 Charles II grants Carolinas to eight proprietors
1664 New Amsterdam seized by the British and renamed New York

Timeline: 1670 to 1801

CANADA

1670 King Charles grants charter to Hudson Bay Company

1684 La Salle explores Mississippi and claims Louisiana for France

1720 French build Fort Toronto as a trading post

1754–63 French and Indian Wars
1759 British defeat the French. The fall of Québec
1760 New France conquered by the British
1763 all French Canadians now became British subjects
1764 construction of Fort Erie at junction of Lake Erie and Niagara River

1774 passage of the Québec Act restores French civil law (now co-existing with English criminal law), French becomes the second language. Catholicism legally recognized and Catholics allowed to hold public office
1775 American revolutionary forces capture Montréal, but are repulsed at Québec

1791 Upper Canada (now Ontario) separated from Québec, which became Lower Canada

1799 St. John's Island renamed after Edward, Duke of Kent

MEXICO

1711 first opera in New World, *La Parténopé* staged in New Spain

1767 Jesuits expelled from New Spain

1786 Spanish King Carlos III appoints 12 intendants to rule New Spain, replacing *acaldes, mayores* and *corregidores*

1789 Count of Revillagigedo appointed Viceroy of New Spain (1789–94)

NORTH AMERICA AND U.S.A.

1682 William Penn founds Pennsylvania

1693 College of William and Mary founded

1704 *Boston News-letter* begins publication

1715 Yamassee War in the Carolinas

1728 Vitus Bering explores Arctic Ocean

1733 colony of Georgia founded

1749–52 Benjamin Franklin experiments with electricity

1763 Treaty of Paris; French Canada and Spanish Florida ceded to Britain. Trans-Mississippi region of Louisiana ceded to Spain

1765 Stamp Act passed. Declaration of Rights and Grievances adopted by Stamp Act Congress in New York
1768 Treaty of Fort Stanwix: Iroquois cede "Ohio country" to the English colonies

1773 Tea Act passed. Boston Tea Party
1774 passage of the "Intolerable Acts". First Continental Congress convened at Philadelphia

1775 American Revolution begins. Battles of Lexington and Concord. George Washington named commander of Continental Forces. Battle of Bunker Hill
1776 Declaration of Independence
1777 Congress adopts Articles of Confederation
1778 Franco-American treaties of alliance. Treaty with Delawares is the first American treaty with Indians. Captain James Cook explores route beyond Bering Strait

1781 British surrender at Yorktown

1783 Treaty of Paris: Britain recognises American Independence
1784 first permanent Russian settlement on Kodiak Island

1787 Congress passes the Northwest Ordinance, accelerating westward expansion
1787–88 new constitution written for the United States

1789 George Washington becomes first President of the United States

1791 first Bank of the United States established in Philadelphia
1792 Stock Exchange opens on Wall Street, New York City
1793 Eli Whitney invents cotton gin

1796 Washington's Farewell Address

1798 Alien and Sedition Acts

1801 Thomas Jefferson inaugurated

Timeline: 1803 to 1867

CANADA

1806 appearance of first newspaper in French, *Le Canadien*

1812–14 War of 1812
1813 Americans blow up Fort York (Toronto) in the War of 1812

1832 completion of the Rideau Canal between Ottawa and Kingston, one of the great engineering feats of the century

1838 first game of baseball played at Beachville, Upper Canada
1839 Lord Durham's report proposes legislative union of Upper and Lower Canada

1841 Act of Union created the United Provinces of Canada

1848 Nova Scotia the first colony of British North America to achieve "responsible government"

1854 opening of Great Northern Railway in Canada
1855 Ottawa becomes Canada's capital by royal decree

1867 Dominion of Canada established. The Bishops of Québec raise a contingent of Papal Zouaves to join the fight against Garibaldi

MEXICO

1804 Act of Consolidation sequesters church funds in New Spain

1808 Viceroy Iturruigaray deposed by Royalists in New Spain

1810–11 revolt by Hidalgo in New Spain

1813 Morelos' Congress issues first Mexican declaration of independence

1821 Mexico wins independence. Rule of Emperor August Iturbide (1822–23)

1824 Estados Unidos Mexicanos organized

1830 Mexican colonization law bans future immigration into Texas

1836 Santa Anna takes Alamo. Massacre at Goliad. Victory by Texans at San Jacinto

1846–48 Mexican War

1857 Mexican federal Constitution sets up unicameral legislature

1861 Britain, France and Spain agree occupation of Mexican coast to settle outstanding claims. After withdrawal of the other powers, the French invade Mexico
1862 Discovery of the first Olmec head

1864 Napoleon III installs Habsburg Archduke Maximilian on Mexican throne

1867 Republican uprising causes collapse of regime of Maximilian. Executed outside Querétaro

NORTH AMERICA AND U.S.A.

1803 Louisiana Purchase. End of the slave trade to American ports

1807–09 Embargo Act closes American ports

1812–14 War of 1812

1817 Seminole wars

1819 United States purchases Florida and Spain

1825 opening of Erie Canal

1827 publication of *Freedom's Journal*, the first black newspaper to be published in the United States

1830 Congress passes the Indian Removal Act. Steam engine loses race with horse

1832 Black Hawk's War. Removal of southeastern tribes (the "Trail of Tears")
1836–44 Texas gains independence from Mexico

1844 electric telegraph opens between Washington and Baltimore
1845 Annexation of Texas
1846–48 Mexican War

1851 publication of Herman Melville's *Moby Dick*. First issue of the *New York Times*

1854 Kansas–Nebraska Act

1857 *Dred Scott* decision invalidates Missouri Compromise

1859 first oil well drilled (Pennsylvania)
1860 Abraham Lincoln elected. South Carolina secedes
1861 southern states declare secession. Civil War begins

1863 Emancipation Proclamation

1865 end of the Civil War; slavery abolished; Lincoln assassinated
1866 first transatlantic cable laid.
1886 Statue of Liberty unveiled
1867 Alaska purchased from Russia

Timeline: 1869 to 1942

CANADA

1869 first Canadian trans-Continental railroad completed. Prince Rupert's Land, Manitoba (1870) and British Columbia (1871) join Canada

1873 creation of the Northwest Mounted Police (the Mounties)—now called the Royal Canadian Mounted Police

1885 completion of Canadian Pacific Railway

1887 Transcontinental Railroad reaches Vancouver, connecting Canada from ocean to ocean

1896 100,000 people swarm into the Yukon during the Klondike goldrush. These were the richest gold deposits ever found

1905 Provinces of Alberta and Saskatchewan established by Autonomy Bill

1914 Canada enters World War I alongside Britain.

1918 60,000 Canadians killed in action in the war
1919 general strike in Winnipeg lasts six weeks

1923 Canada secures right to conduct its own foreign policy at Imperial Conference

1926 constitutional independence asserted at Imperial Conference

1930 depression and mass unemployment hit Canada. By 1933 a quarter of the workforce is unemployed

1939 Canada enters World War II with Britain

1942 72 per cent of Québec voters oppose conscription. 80 per cent of English-speaking Canadians vote "yes". In a commando raid on Dieppe in August, 66 per cent of the Canadian forces were casualties

MEXICO

1876 Porfirio Díaz gains control of Mexico (to 1911)

1911 Mexican Revolution begins

1916 Pancho Villa's men sack Columbus, New Mexico, killing 18 Americans
1917 Querétaro Congress drafts new anti-clerical Constitution

1919 assassination of Emiliano Zapata

1935 Cárdenas becomes President of Mexico: land redistribution and (1938) nationalization of oil

NORTH AMERICA AND U.S.A.

1870 John D. Rockefeller founds the Standard Oil Company of Ohio

1872 Modoc War

1874 telephone patented by Alexander Graham Bell
1876 Battle of Little Big Horn
1877 Chief Joseph's War

1883 opening of the Brooklyn Bridge

1886 United States overtakes Britain in steel output. Statue of Liberty unveiled

1889 Oklahoma land rush
1890 Battle of Wounded Knee. Last armed resistance to Indian removal to reservations

1893 Duryea brothers demonstrate first American gasoline-powered automobile

1898 Spanish–American War: U.S. annexes Guam, Puerto Rico and Philippine Islands

1901 Marconi's first transatlantic radiotelegraphy message. Incorporation of United States Steel Company

1903 Panama Canal Zone ceded to United States. First powered flight by Wright brothers

1910 foundation in New York City of the National Association for the Advancement of Colored People

1914 Henry Ford develops conveyor belt car assembly

1917 United States declares war on Central Powers
1918 President Woodrow Wilson announces "Fourteen Points"
1919 Senate rejects League of Nations Treaty. Prohibition ratified. First transatlantic flight by Alcock and Brown
1920 first radio broadcasts. Female suffrage ratified

1927 Lindbergh's solo flight across Atlantic

1929 Wall Street Crash precipitates world Depression

1931 dedication of the Empire State Building in New York City

1933 bank crisis. "New Deal" introduced by President Franklin D. Roosevelt

1935 Social Security Act passed

1936 Pan-American Congress: U.S. proclaims good neighbour policy

1941 U.S. begins "lend-lease" to Britain. Japanese attack on Pearl Harbor precipitates entry into World War II
1942 Internment of 110,000 Japanese on West Coast

Timeline: 1945 to 1994

CANADA	MEXICO	NORTH AMERICA AND U.S.A.
1945 nearly 42,000 Canadian dead or missing in the war		1945 first atom bomb exploded. Allies victorious. United Nations establishes headquarters in New York City
1947 enormous deposits of oil discovered in Canada		1947 Truman Doctrine and the Marshall Plan. Construction of first Levittown 1948 Organization of American States established
1949 Newfoundland becomes Canada's 10th province		1949 North Atlantic Treaty Organization approved 1950 Korean War begins 1951 first nuclear power stations opened
1952 first native Canadian, the Right Honorable Vincent Massey, appointed governor-general		1952 contraceptive pill introduced
		1954 in *Brown versus Board of Education* the Supreme Court declares racially segregated schools unconstitutional
1959 opening of the St. Lawrence Seaway turns Toronto into a major seaport		1959 Cuban revolution
		1961 increased U.S. involvement in Vietnam. Bay of Pigs invasion of Cuba fails 1962 Cuban missile crisis 1963 President J.F. Kennedy assassinated 1964 Civil Rights Bill inaugurates President Johnson's "Great Society" programme 1965 Johnson "escalates" Vietnam War. First student teach-in (University of Michigan). Race riots in Watts, Los Angeles
1967 Canada's 100th anniversary. Montréal hosts the World Expo. The founding of the National Liberation Front of Québec. De Gaulle's "Long live free Québec" speech		1966 *Miranda v. Arizona* decision requires criminal suspects in the United States to be informed of their rights
1968 Parti Québeçois founded by Réne Lévesque (the beginning of the separatist movement) 1970 kidnap and murder of Cabinet Minister Pierre Laporte. War Measures Act imposed on Québec province	1968 hundreds of students killed when police open fire on a demonstration in Tlatelolco	1968 assassination of Martin Luther King 1969 astronaut Neil Armstrong is the first man on the moon 1970 first "Earth Day" in the United States 1971 policy of détente with USSR and China initiated by President Nixon 1972 arrests at Watergate complex in Washington precipitates political crisis. Signing of SALT I Treaty 1973 OPEC oil embargo and price rise causes recession. Military withdrawn from Vietnam. *Roe v. Wade* decision invalidates all laws prohibiting abortion
1974 French adopted as the official language of Québec		1974 President Nixon resigns in aftermath of Watergate scandal. 1975 final withdrawal of the United States from Vietnam. Communist victory
1976 Olympic Games held at Montréal		
	1978 Aztec remains of the "Great Temple" found in Mexico City	1978 Camp David accord signed by Sadat and Begin 1979 accident at Three Mile Island nuclear plant in Pennsylvania
1980 Québec votes "non" to separatism		
1982 new constitution, signed by Queen Elizabeth II, makes Canadians masters in their own house (but not endorsed by Québec)		1982 AIDS cases appear in United States
	1985 earthquake in Mexico City	
1988 Calgary hosts the Winter Olympics 1989 Canada-U.S. Free Trade agreement eliminates all tariffs on goods of national origin moving between the two countries 1990 struggle between Mohawk tribes and the Sûrete Québec police force highlights issues of land disputes and Native rights in Canada		1990 census reveals population of U.S. to be 248,709,873, a 9.9 per cent increase since the 1980 census. 1991 Gulf War expels Iraqi troops occupying Kuwait
1993 signing of North American Free Trade Agreement (NAFTA)	1993 signing of North American Free Trade Agreement (NAFTA) 1994 peasant uprising in Chiapas	1993 signing of North American Free Trade Agreement (NAFTA)

I: The First Peoples

From the appearance of the first hunters to the flowering of complex urban centres like Tenochtitlán, the development of civilization gradually unfolded in North America.

As they crossed the Bering land bridge into North America, those first small bands of hunters, carrying little more than weapons (perhaps a fire-hardened stick or pointed stone), of course brought their culture with them. It was a culture of the hunt and of the harsh Asian winters, and what they knew was that the deities to which they prayed, and the animals they followed and killed, enabled them to survive. The bands were small, and had come from diverse Asian-Mongolian tribes.

They spread across the continent no faster than the animals they hunted. Finding their unerring way to water and food, the animals over time crossed the flat, scoured landscape where the glaciers had been, and penetrated the deserts, dry plains and thick forests. No memory remained of Asia, or of the

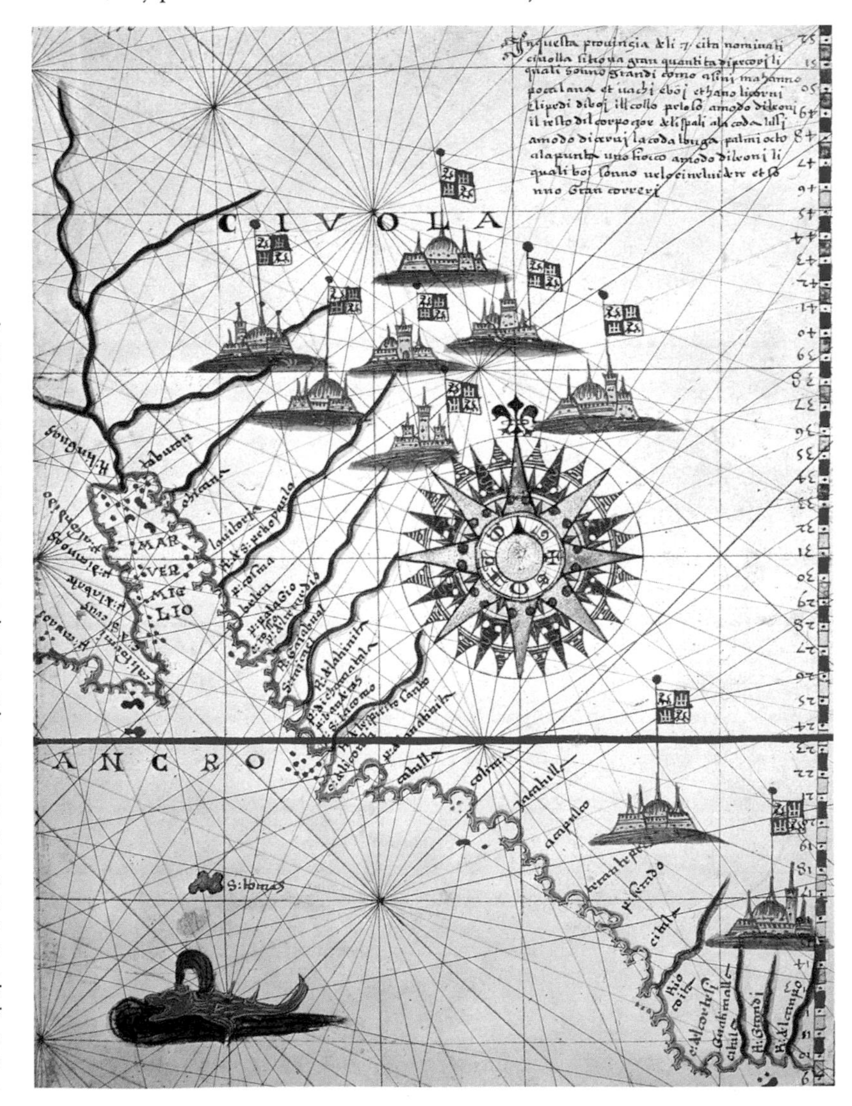

Under repeated pressure of Moorish invasion and conquest, the legend of the Seven Cities of Cibola in medieval Spain told of seven bishops and their followers who set sail to found new cities in unknown places in the west. The Seven Cities were located variously in the New World in places just tantalizingly beyond the reach of Spanish explorers. Any decorated tool, piece of jewellery or tanned leather encountered by Spaniards in the transactions with Native Americans was taken as proof of the existence of a wealthy and highly civilized people to the west. Coronado's expedition of 1540–42 chased these chimeras from Compostella, Mexico, as far as Kansas. The open prairies and modest adobe villages he encountered persuaded no one that infinite wealth was to be found just over the next hill. Juan de Martinez's map of 1578 (right) *embodied one of the most enduring fantasies in the Spanish encounter with North America.*

Located on the plaza at New Chichén Itza, the Castillo was dedicated to the god Kukulkan, and was connected to the Sacred Cenote by a raised causeway. It has nine terraces and stairways on all four sides. It is the tallest ancient building in this part of Yucatán, and encloses a smaller identical structure where a Chacmool was found. The Chacmool, an altar probably used to hold offerings, confirms the ceremonial purpose of the Castillo.

land bridge which seemed no different from the rest of the large territory temporarily transformed into dry land through which they moved during the Ice Age. The hunters did not know when the ice began to recede, or when the land bridge was submerged. But it served to isolate the hunters and their prey for thousands of years.

There is perhaps no evolutionary step more astounding than that which separates these hunter-gatherers from the inhabitants of the coastal swamps of southern Mexico, the Olmecs, who by 1200 BC had evolved a system of carved hieroglyphs and had a form of calendar. (They are now called "Olmecs" because an archaeologist in 1929 decided on the name, now widely used. We know nothing of what the Olmec called themselves). Civilization had evolved stage by stage from the most primitive of hunting techniques to the evolution of a heavy projectile point which extended the range and effectiveness of their hunt. The improvements in techniques for preparing the projectile point are themselves important markers "Clovis-

First came the fur traders, the coureurs de bois, *with their trinkets and brandy. Then the garrisons, with drunken troops who stole from the Indians, gave them alcohol, and did their best to turn every European outpost into a brothel. And at last came the priests, who denounced the corruption of native life and demanded that the Indians be kept near the missions. The French knew that if they failed to provide brandy for the Indians, the Dutch and the British would, and the fur trade would be lost. When the priests were not denouncing the traffic in brandy, and trying to save souls, they sent striking pen and ink drawings to France of the curious appearance of the* sauvages, *their light canoes* (left) *and strange dwellings* (right).

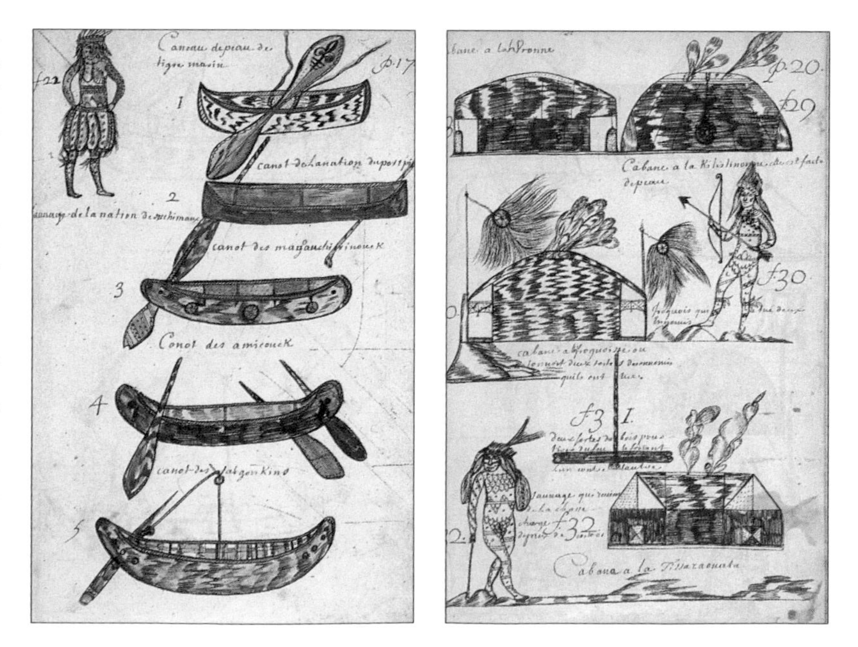

type" fluted points appear between 7000 and 12,000 years ago, overlapping the "Folsom-type", used by hunters of the huge mammoths between 10,000 and 11,000 years ago, and the "Cascade" point, used by hunters spreading down the Pacific Northwest from Alaska from 9500 to 12,000 or 13,000 years ago. The accidental discovery of archaeological sites have given names to epochs of human development.

Mesoamerican weapons and tools were made of chipped stone. Some, as in this example of a ceremonial knife made of obsidian, were highly decorated. Such an object was used in the human sacrifices which were central to Mesoamerican religion. The heart would be removed from a sacrificial victim before being displayed and then eaten. The skulls of the sacrificed were held on large racks nearby.

Agriculture, beginning early in the most fertile and well-watered river valleys, slowly spread among the hunters, at first supplementing the irregularity of their food supply. In time, when the large mammals disappeared, and the hunters pursued smaller animals, the gathering of wild plants and roots evolved into a recognizable agriculture, and provided a more abundant food source. Populations grew, and permanent settlements were created near good water and fertile soil. Women became full-time agriculturalists long before men. They created food-grinding implements to prepare the wild maize which they gathered, and wove the reed baskets or fired the clay pots to prepare and store food. There were no ploughs, or draft animals to provide manure to fertilize the soil. A slash-and-burn agriculture, wasteful of land, proved sufficient to feed a growing population. When the gods of the soil turned their faces away, the people prayed, offered sacrifices, and moved on.

Evidence exists of village life in Tehuacán 5300 years ago. To the north, in the forests which covered much of North America, a "Woodland Culture" began to appear about 3000 years ago. With settlements characterized by burial mounds and circular breastworks, it is clear that Woodland cultures, such as the Adena and Hopewellian, were sedentary. The discovery of goods

Hopewell artisans also worked in clay. The kneeling figurine (below) *with a topknot, may be a shaman. Found in an unknown location in Ohio, the shell head* (below right) *suggests the presence of ritual scarification and facial decoration in Mississippian southern culture. Weeping eyes and the topknot appear in other artefacts.*

Jaina is a small limestone island off the west coast of Campeche. The many Maya tombs found at Jaina indicate its use as a sacred site for burials. The hollow figurines found in Jaina tombs are highly individualized but they may represent deities. Each has the mouthpiece of a whistle at the back. The figurines are small and highly portable. Many were looted from Jaina and sold to collectors and museums.

originating from great distances suggests the existence of trade, and of the development of woven fabric, pottery and other specialized crafts used to trade for exotic goods from afar. It is possible that the continent was laced with trade routes, laboriously traced by foot (in the absence of the horse). The need for ever finer goods to adorn priests and great chieftains and to deposit in funeral mounds, was a powerful incentive to cultural production and economic activity.

Olmec culture stands apart from the archaeological evidence provided by artefacts from the territory which became the United States. Olmec craftsmanship exhibits far higher levels of technical skill, and more realistic artistry. There is no work more monumental on the continent than the great helmeted Olmec heads, found in 1862 deep in the jungle. The appar-

The presence of copper in Hopewell provided a soft, malleable material for representational figurines. A raven or crow cut from native sheet copper, (above left), *and a mica claw of a bird of prey* (above right), *were found at locations in Ohio.*

ently "negroid" facial characteristics of the heads (and the presence of more obviously "Asiatic" features on other examples of Olmec art) has encouraged much unfounded theorizing about possible African migrations to Mesoamerica. No evidence whatsoever exists to suggest that either Asian, African or any other peoples made appearances in ancient North America. The Olmec may simply have been a racially mixed civilization. There is little evidence other than artefacts to provide further clues.

The Olmec lived in the region to the west of the River Tonalá, in the states of southern Veracruz and Tabasco. Evidence of their presence has been found at three sites. The oldest, at San Lorenzo, was occupied between 1200 and 900 BC. Important construction at La Venta, on a remote island in the midst of swamps, was ceremoniously demolished around 500–400 BC. On the western edge of Olmec settlement was Tres Zapotes, where Olmec culture persisted until as late as 100 BC.

The lineaments of the later Mesoamerican civilizations are present in Olmec life, and the picture we derive from the archaeological evidence is widely applicable. Theirs was a society based on the cultivation of maize,

The Ball Court at Copán is among the best-preserved of Mayan sites. Located on a branch of the Rio Motagua in lowland Honduras, Copán was a provincial capital of some importance. When the site was first explored in 1839 it was found to contain temple-pyramids, altars, stelae, three-dimensional sculpture and the Ball Court.

which made up the largest part of the Olmec diet. The vast majority of people in the Olmec territory would have worked at slash-and-burn agriculture, surrendering their surplus to support the existence of artisans who were producing the exquisite jade carved figurines, many of which were variations upon the jaguar, the dominant motif in Olmec art and the central cult in their religion. There would have been crews of builders, stone-cutters, masons, sculptors and labourers (probably slaves) to haul heavy material from great distances to create the ceremonial centres. Traders—perhaps warrior-traders—would have been needed to carry the Olmec artefacts which have been found as far south as Guatemala and as far north as the Valley of Mexico. And, to tend the cult of the jaguar who was a man (and a man who was a jaguar), which was central to the rites of sacrifice upon which Mesoamerican religion was based, a priesthood was needed. Priests kept records of the festivals and days of prayer, and marked the circuit of the year in their hieroglyphic writing and calendar. And there would have been a warrior class who defended the realm against foreign enemies, captured slaves needed for labour and sacrifice, and imposed obedience within the Olmec world.

Mesoamerican civilization began with the Olmecs. Other universal cultures followed: the Maya; the urban centre of Teotihuacán; the Toltec settlement at Tula, which lasted barely 200 years; and then, in 1428, the establishment of the Aztec Empire which was finally destroyed in 1521. Scarcely a century separates the last Olmec from the first Maya art forms. The Toltecs brought their I-shaped ball courts with them when they invaded Yucatán. Their culture survived at Culhuacán to provide a link with the Aztecs. The stunning discontinuities of Mesoamerican history have no parallel elsewhere on the continent. Nor a scene as poignant as that which greeted the Spanish when they encountered the great ruined sites of Maya civilization, inhabited by a people who could no longer decipher their own religious texts and knew nothing of their people's history. A scene worthy of Gibbon, perhaps.

Pipe smoking, (below left), and a sun dance ceremony, probably Cheyenne, recorded on Indian cloth, (below right), were central to rituals practiced by Plains Indians.

Hunters Spread Across a Continent

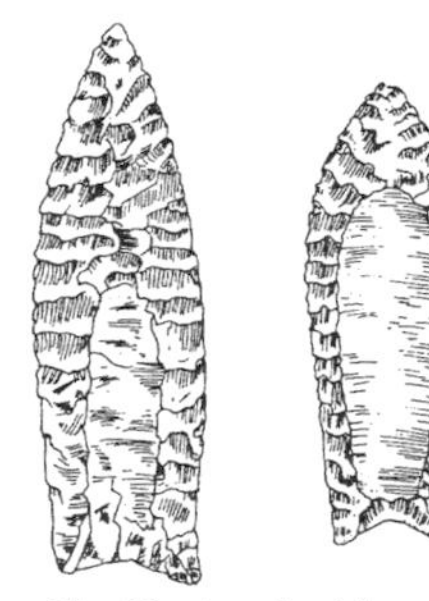

The Clovis point (above left) *is named after the site in New Mexico where it was first found. This finely flaked spear head was used from around 11,000 BC to hunt large mammals such as mammoths. After these became extinct around 10,000 BC, it appears to have been superseded by the lighter Folsom point* (above right)*, which has been found with the skeletons of long-horned bison.*

The Pleistocene geological era that began about 2.5 million years ago and lasted until around 8,000 BC, corresponded to the last great ice age. That ice age was not uninterrupted, however; there were periods when the ice advanced and others (interstadials) when it retreated. When the ice advanced, the great glaciers held so much of the planet's water that the sea level dropped and a land bridge formed across what is now the Bering Strait between Siberia and Alaska. It was during one of these periods—there is still controversy about which one—that hunters first crossed into North America.

Ancient bridges of land spanning the continents enabled migrants to cross into North America from Asia and north China via Siberia.

The gradual retreat of the great ice sheet of the Pleistocene Age, which covered much of North America, had largely been completed 10,000 years ago. It was during the last phase of the Pleistocene when the first groups of Arctic hunters crossed into the North American continent from Siberia. Even a temporary land route on broken ice fields, or by way of a shorter sea voyage between the small islands which dot the Bering Sea, opened up vast new spaces in which small packs of hunters could follow their prey. The hunters probably came from northern China, but over time other groups of Asiatic hunters made their way across the bridge, carrying little food or other materials. The crossing of the land bridge probably occurred between 30,000 and 40,000 years ago.

It may have been hunger that led hunters to venture from Asia in search of food, unaware that they had crossed onto a new continent. The descendants of the first explorers moved gradually southward, into a continent emerging from the great ice sheet which was gradually retreating towards the Arctic. Large mammals and mastodons provided the hunters with sustenance, and when they grew too few in number the small bands of hunters moved on, looking for new prey. They also fed off roots, plants and berries. The hunter-gatherers of the Pleistocene fashioned clothing, wove baskets and used nets for fishing. The bands remained small because they were totally dependent upon local resources. When food became seasonally scarce, an increase in numbers was a threat to their chances of survival. The need to separate into smaller, more mobile packs, was frequently necessary. Within several generations bands which were thus divided became highly differentiated in language and observance. Until they could command more control over their sustenance, there was little chance of expanding their numbers, or developing the skills which would enable permanent settlements to be established.

The first settlements of hunter-gatherers in Mexico have been dated to around 20,000 BC. At Tepexpan, 25 miles north of Mexico City, human remains have been found which date from 9000 BC.

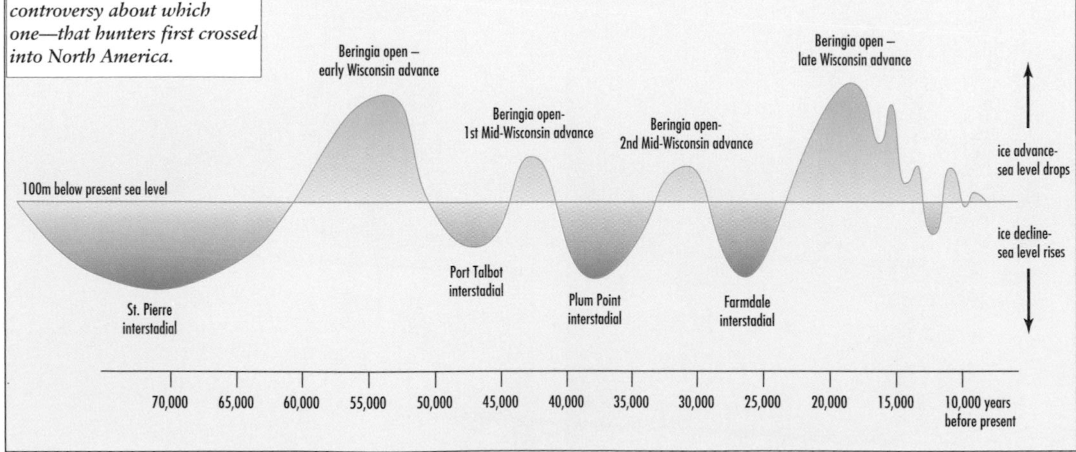

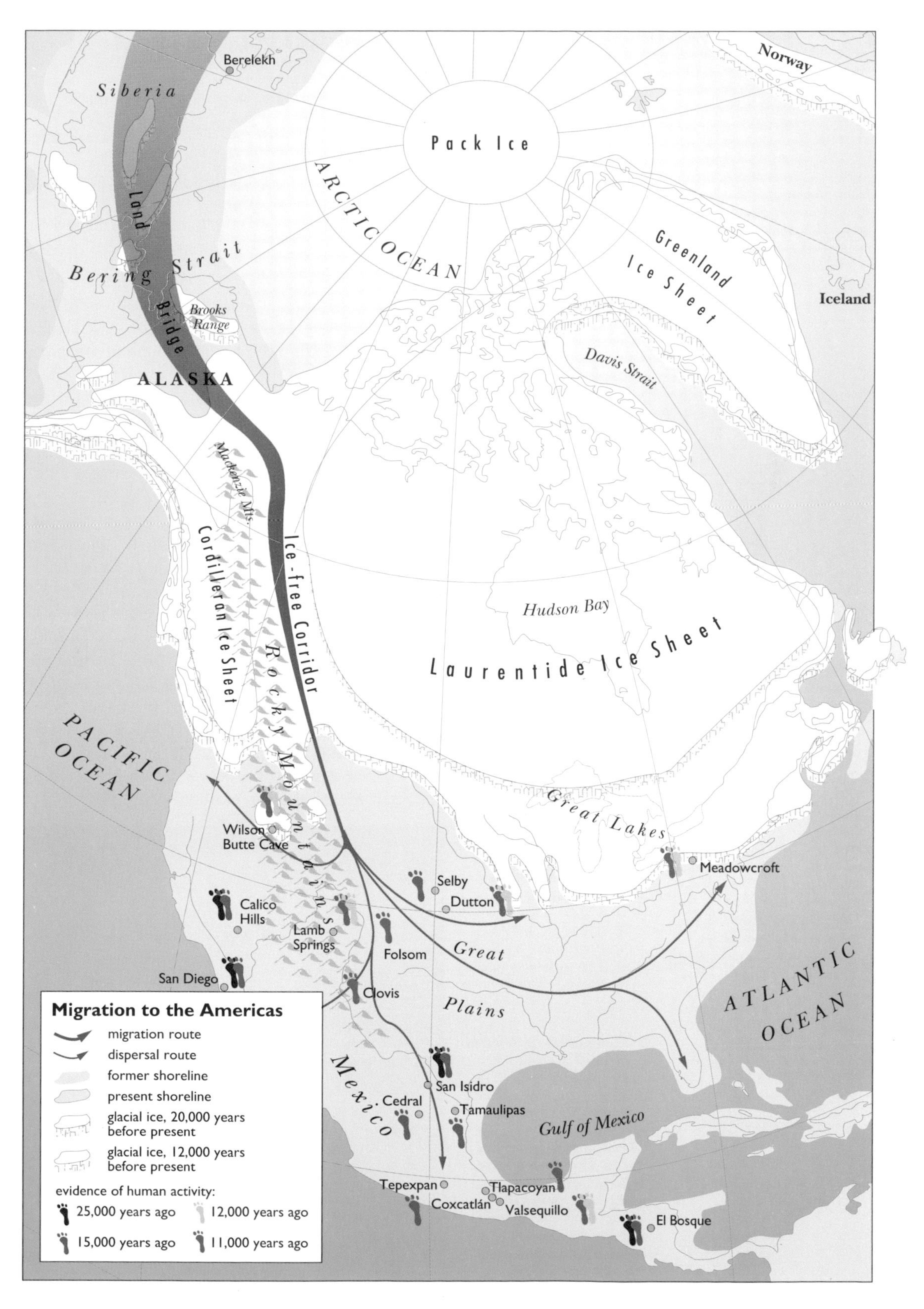

Migration to the Americas
migration route
dispersal route
former shoreline
present shoreline
glacial ice, 20,000 years before present
glacial ice, 12,000 years before present
evidence of human activity:
25,000 years ago
12,000 years ago
15,000 years ago
11,000 years ago
Berelekh
Siberia
Pack Ice
ARCTIC OCEAN
Norway
Land Bridge
Bering Strait
Brooks Range
ALASKA
Greenland Ice Sheet
Iceland
Davis Strait
Mackenzie Mts.
Cordilleran Ice Sheet
Ice-free Corridor
Rocky Mountains
Hudson Bay
Laurentide Ice Sheet
PACIFIC OCEAN
Great Lakes
Wilson Butte Cave
Meadowcroft
Selby
Dutton
Calico Hills
Lamb Springs
Folsom
Great Plains
San Diego
Clovis
ATLANTIC OCEAN
Mexico
San Isidro
Cedral
Tamaulipas
Gulf of Mexico
Tepexpan
Tlapacoyan
Coxcatlán
Valsequillo
El Bosque

First Civilizations: Olmec and Maya

Independent civilization in North America began as early as 1200 BC among the aboriginal inhabitants of southern Veracruz and the swampy areas of western Tabasco.

The site of an Olmec ceremonial centre has been found by archaeologists at San Lorenzo, on a branch of the Coatzacoalcos River in Veracruz, which was inhabited as early as 1200 BC. At a time when Nebuchadnezzar reigned in Babylon, the Olmec had evolved a means of representation with carved hieroglyphs, and had invented a form of calendar.

The Olmec cult of the jaguar flourished at La Venta on a remote island in the Tonalá River which was occupied from 900 to about 600 BC. The jaguar cult, combining human and jaguar features, was based upon ritual offerings to the rain gods and gods of fertility. The Olmec custom of burying offerings to the gods, and of smashing ceremonial altars, culminated in the abandonment of whole sites like La Venta, apparently without external threat. Huge stone heads, some weighing up to 40 tons, representing helmeted warriors, hint at the power and organization of a civilization which had disappeared into the jungle some 2000 years ago.

Located near the centre of Mesoamerican ceremonial complexes, the ball game, ollama, *was clearly much more than a game. There were 11 such courts at Tajín alone. Using a heavy solid rubber ball, the game was played by teams of seven men wearing protective belts made of wood and leather, and pads on knee and upper arm. The aim was to keep possession of the ball with hip and elbow. Scoring was achieved by getting the ball through a stone hoop placed vertically on the wall. The losing team forfeited their outfits to the winners, and some ball court friezes show the losing captain being decapitated.*

Similar stirrings occurred in small horticultural settlements of the Maya on the Yucatán peninsula. From AD 250 until their fall in AD 900, the Maya constructed a world of masonry temples, palaces, plazas and altars which were embellished with ornate carvings, and decorated with friezes, frescoes and murals. Polychrome pottery of great beauty has been found at Maya sites. The remoteness of their location, and the fierceness of the Maya people, largely kept them independent of the other substantial Classic civilization centred around Teotihuacán in the central valley of Mexico, 25 miles northeast of Mexico City. Of the two states, Teotihuacán covered the largest area, and engaged in the most widespread commerce. The Maya palace at Palenque was 300 feet long and 240 feet wide. Temple IV at Tikal was 229 feet high. The civilization of the Maya was monumental and complex, representing the high point of Classic Mesoamerican culture.

The colossal basalt helmeted head found at La Venta weighs some 40 tons. Believed to portray secular rulers, the heads emphasize the grandeur and power of Olmec warriors and rulers. The human labour involved in the quarrying and transporting from a distant location such a large stone suggests a high level of social organization—and a large number of slaves.

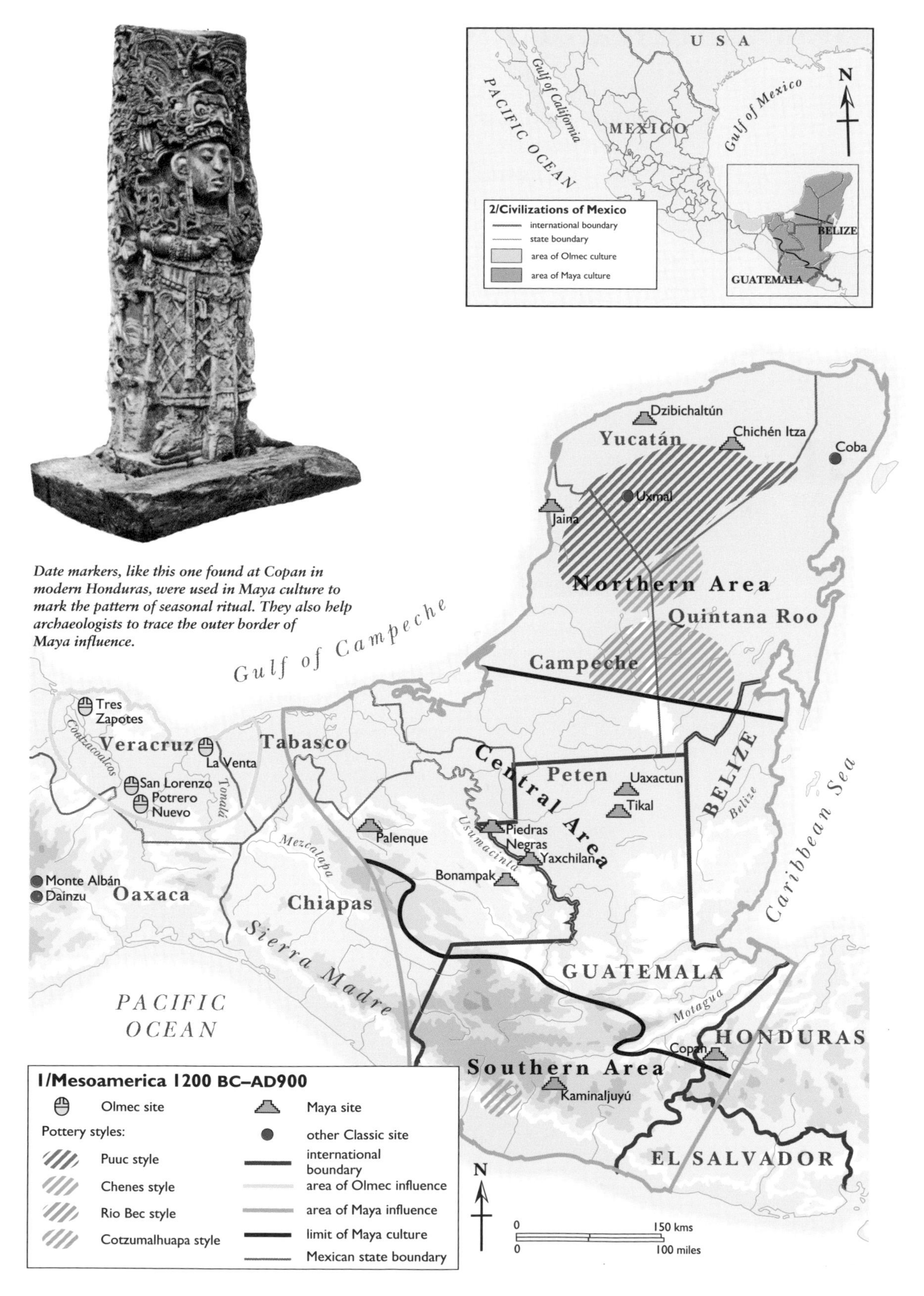

Date markers, like this one found at Copan in modern Honduras, were used in Maya culture to mark the pattern of seasonal ritual. They also help archaeologists to trace the outer border of Maya influence.

Mesoamerica: Toltecs and Aztecs

The Post-Classic Period (950–1519) was dominated by two far-flung empires based upon the great urban centres of Tula and Tenochtitlán.

"You will be limited by nothing... whatever your greeds are you will be satisfied, you will take women where and when you please... you will receive gifts of everything—the best food, the greatest ease, fragrance, the Flower, tobacco, song, everything...." The promise of the god Huitzilopochtli to the Aztecs if they became warriors.

By AD 1000, the Toltecs had transformed the village of Tula into the most powerful city in middle Mexico. In its heyday, from AD 950 to 1150, Tula was dominated by temples built upon vast pyramids. Toltec goods were highly esteemed and the production of pottery for everyday use involved as many as a third of the inhabitants of Tula. The city was a hive of family compounds producing crafts, armaments and pottery. But economic decline exacerbated internal conflicts. The heterogeneous population had never been socially integrated, and by 1200 the Toltec empire collapsed, the statues of the deities were smashed, and the Toltec city was in ruins.

The Aztec people alone succeeded in recreating a great pan-Mesoamerican empire out of the disorder which followed Tula's collapse. Speaking the language of Tula—Nahuatl—and claiming to be the successors of the Toltecs, the Aztecs wandered throughout central Mexico, a primitive people without power. It was not until they settled near Chapultepec that they began to absorb the culture of their neighbours, the Tepanecs. The payment of tribute, the waging of war on behalf of an overlord, toughened the Aztec and perfected their militaristic social order.

Settling on the island of Tenochtitlán, the Aztec rose to pre-eminence within the Valley of Mexico. By 1520, Tenochtitlán had a population of 360,000, making it one of the world's three largest cities. They waged incessant war to provide human sacrifices to the gods. No other Mesoamerican people believed that the ritual murder of slaves, captives and children on such a scale was necessary for the very survival of the universe.

From 1430 the Aztec received tribute from the Gulf to the Pacific. By the time Cortés arrived, Montezuma II, the *uei tlatoani*, Chief Speaker, enjoyed near-divine status. Priests and a military oligarchy presided over a perfected militaristic theocracy.

With human sacrifice so central to Mesoamerican culture, war was an honoured way to obtain prisoners for sacrifice. The more noble and exalted the prisoner, the more elaborate were the rituals attending the sacrifice. To the immediate left of the ceremonial pyramid warriors cheer, while a great prince observes the bloody decapitation and dismemberment of a prisoner sacrificed to the Sun God. Under Ahuízotl in 1487, as many as 20,000 hearts were ripped out, over four days, to honour the god Huitzilopochtli.

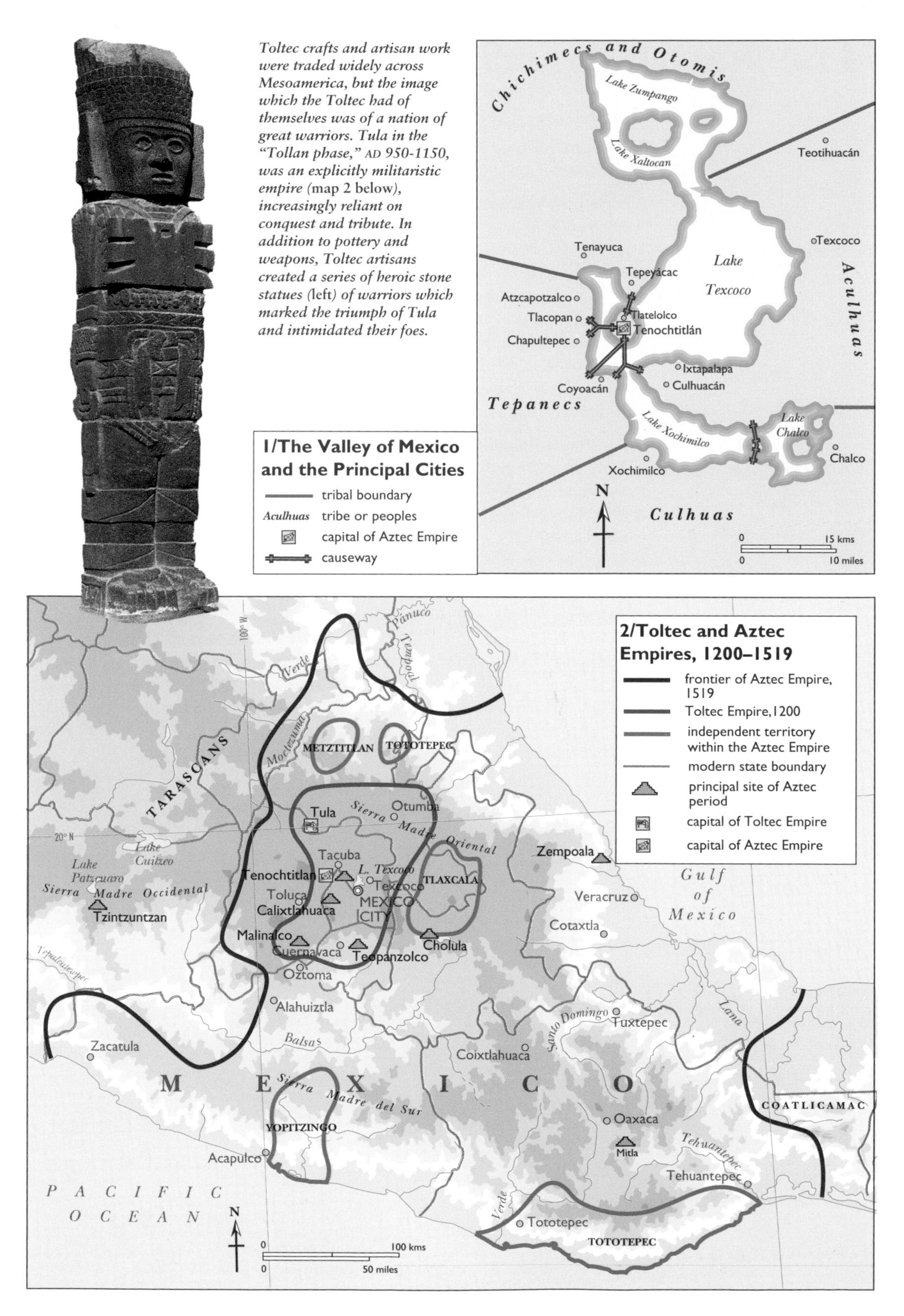

Toltec crafts and artisan work were traded widely across Mesoamerica, but the image which the Toltec had of themselves was of a nation of great warriors. Tula in the "Tollan phase," AD *950-1150, was an explicitly militaristic empire* (map 2 below), *increasingly reliant on conquest and tribute. In addition to pottery and weapons, Toltec artisans created a series of heroic stone statues* (left) *of warriors which marked the triumph of Tula and intimidated their foes.*

The Native Americans

The ancestors of the neolithic hunter who spread across the continent adapted their way of life to the conditions they encountered.

"Three women go in, one of them breaks the corn, the next grinds it, and the third grinds it again. They take off their shoes, do up their hair, shake their clothes, and cover their heads. A man sits at the door playing on a fife while they grind, moving the stones to the music and singing together."
From *The Narrative of the Expedition of Coronado,* by Pedro de Castaneda, 1540

Agriculture was initiated in North America in different places between 7,000 and 3,000 years ago, but made no sudden revolution in the slow pattern of neolithic life. The availability of game, fish and the many foodstuffs of woodland areas determined an economic system based upon subsistence production and trade. The cultivation of maize, which became a major foodstuff wherever it could be grown, was accompanied by the discovery of new foods like the potato, pepper and tomato. The vanilla and cacao bean, both produced in the tropics from fermented and dried pod seeds, were used for trade; pottery and crafted and decorated objects appear in burial mounds far from their place of origin.

There was no pan-Indian identity or culture, no common language. When Columbus arrived in the New World in 1492, it has been estimated that the native inhabitants spoke 2200 different languages (many with substantial regional variations). Indian life was strictly governed by local conditions and tribal culture. Where language or cultural ties with other tribes existed, they were pragmatically perceived as loose confederacies—temporary alliances for plunder or trade.

The Indian peoples had no notion of private property like that assumed by Europeans. They occupied land by tradition, not legal title. Maps with clearly defined Huron or Pawnee territory are less helpful than those which suggest language groups or "culture areas". Tribes repeatedly moved throughout their history, and when the horse was encountered in the 16th century (brought from Spain by the Conquistadors), movement itself became a way of life.

There was a need for communication between tribes with mutually incomprehensible languages. For this purpose, *wampum,* made from pierced seashells, was a unit of value, and the *calumet,* the French name for a long-stemmed peace pipe, accompanied emissaries or messengers as a sign of peaceful intention. The *calumet* also played a part in tribal councils and certain religious rites.

The Iroquoian and Huron longhouse, which would be arranged around an open space at the centre of the settlement, could be up to 50 yards long, and 12 or 15 yards wide. Made with flexible saplings covered with sheets of bark, the longhouse was light, well-insulated and portable. Each contained smoke holes, and could have several entrances. A number of families would have lived in each longhouse.

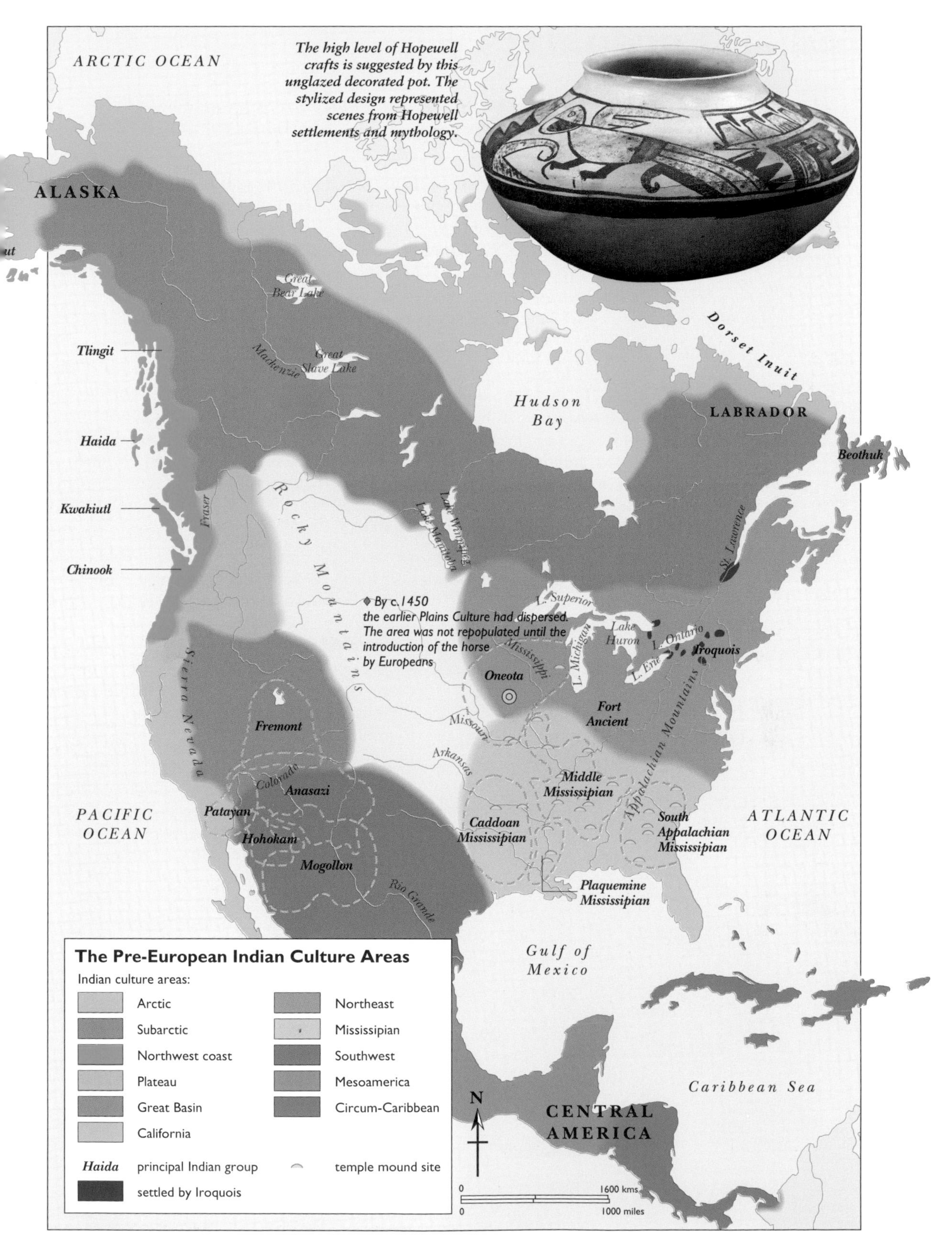

The high level of Hopewell crafts is suggested by this unglazed decorated pot. The stylized design represented scenes from Hopewell settlements and mythology.

Indian Encounters

Early contacts between Indians and Europeans, made lethal by disease, brought the alien cultures together in mutual incomprehension.

"... not a month of this summer [1651] passed without our role of slain being marked in red at the hands of the Iroquois."
Fr. Francois Dollier de Casson, *A History of Montréal 1640–1672*

Assuming in 1492 that he had arrived in the Indies—that is, in the Far East—Columbus named the aboriginal peoples he encountered *los Indios.* Having long outlasted the French and Dutch equivalents ("*sauvages*", "*Wilden*"—savages), we are largely stuck with it. The name "Indian" obscures the remarkable diversity of peoples who inhabited the continent when the Europeans arrived.

Amidst the Babel-like confusion of languages and dialects spoken by the Indians, there was no single word to describe themselves. The Delawares, living on the bank of the river named after them, knew themselves as "Lenni Lenâpé", or "common people". When they encountered the Dutch, who had established a trading settlement on Manhattan, they called the strangers "Swannakans". Was it more difficult for illiterate Dutchmen to grasp the linguistic and tribal differences between Mohican, Mohegan and Mohawk, or for the Indians to see all the things which differentiated the French from the British, Dutch and Swedes, Protestant and Catholic, who had all intruded into their land?

To European ears, the very names of the natives are confusing. Manhattan was named after the Manates, a Munsee tribe of the Wappinger confederacy, who were also referred to as Manatuns, Manhates, Manhatesen and Manhattans. Terms such as "Algonquin" and "Iroquoian" were European summary descriptions of what are taken to be linguistically related peoples—who could speak mutually incomprehensible languages and be long-standing rivals.

The Europeans who made contact with the Indians learned at least the rudiments of their language. Samuel de Champlain sent a young man, Etienne Brulé, to live among the Hurons in 1609. A year later, Brulé reappeared, dressed in Indian clothes and speaking Huron. Brulé became the first major French overland explorer, but his countrymen were suspicious of any European who had "gone native". It was a fear well-justified: in 1629 Brulé went over to the side of Scottish traders who had seized a French trading post. The French *coureurs de*

Frontispiece illustration from Daniel Clarke Sanders' A History of the Indian Wars with the First Settlers of the United States *(1812). This book aroused bitter criticism because of the author's condemnation of colonial bigotry and cruelty to the natives. Sanders (1768–1850) was the first president of the University of Vermont.*

Right: *James Isham's sketch of Indians on a beaver hunt, drawn for the Hudson Bay Company, was made about 1740.*

Above right: *George Catlin (1796–1872) abandoned a career as a portrait painter in Philadelphia for a journey to the West in 1832 to paint the Plains Indians. "Oh how I love a people who don't live for the love of money!" His* North American Indians *(1832–39) celebrated the intense individualism of the Indians and the drama of "Indians Catching Wild Horses on the Plains".*
Above left: *Osceola was a Seminole prisoner of war in Ft. Moultrie, South Carolina, where Catlin painted his portrait. The post surgeon told Catlin that Osceola was "grieving with a broken spirit" and was not expected to live many weeks in captivity.*

bois ("runners of the woods", or "bushlopers"), who were the Europeans most likely to have detailed knowledge of tribal customs and languages, were the carriers and traders who linked the Québec fur merchants to the network of trading posts located deep in Indian territory.

Ancient tribal conflicts among the Indians were exploited by Europeans, and vice versa, in the best tradition of divide and conquer. The Dutch formed lucrative trading relations with the Iroquois while the French allied with the Algonquins and Huron. Contacts with Europeans had disturbed traditional Indian cultures, and soon proved lethal. Cartier kidnapped a group of Indians when he returned to France after his first winter in Québec in the 1530s. Columbus organized the seizure of 1500 natives in 1594, and had 500 sent to Seville to be sold in the slave market.

Well-merited Indian distrust of Europeans was balanced by a strong and growing desire for European trade goods, metal-work and guns. With trade came European diseases. A Spanish dream of vast native labouring populations was destroyed in a generation by the terrible effects of disease, especially when combined with the countless abuses of *encomienda* labour. A census of Hispaniola in 1498 revealed the existence of 1,500,000 tribute-paying Indians. Within a few decades there were effectively none left. The Iroquoian Indians Cartier encountered on the St. Lawrence in the 1530s were gone by 1600. There had been a terrifying and inexplicable dying off. By the time Champlain established his permanent settlement at Québec in 1608, disease, and attacks by raiding parties from other tribes, had left the St. Lawrence valley an empty land.

II: European Intervention

Warfare against the Indians was followed by warfare against each other as European powers sought to extend their territory in the New World.

"There is enough land. We should be good neighbors." Sir John Harvey, Governor of Virginia, welcoming a Dutch visitor on his way to New Amsterdam, March 1633

After its fall in 1521, the Aztec Empire was of interest primarily to a small number of Spanish priests hoping to bring the solace of Christianity to the natives. Priests were among the only Spaniards who troubled to learn something of the Aztec tongue, Náhuatl, and who sought to preserve the records and memories of a defeated and demoralized people. Among their compilations was the *Annals of Tlatelolco*, written in 1528 in the aftermath of Cortés' victory. The *Annals* contain an Aztec lament for their own destruction. It is a unique and terrible vision of the end of the world:

> And all this happened among us... .
> On the roads lie broken shafts and torn hair,
> houses are roofless, homes are stained red,
> worms swarm in the streets, walls are spattered with brains.
> The water is reddish, like dyed water;
> we drink it so, we even drink brine...
> Whatever was still alive was kept between shields,
> like precious treasure, between shields, until it was eaten.

As the final verse suggests, during the siege of Tenochtitlán, food became so precious that dogs (which had formed part of the Aztec diet for many decades), and also perhaps slaves—no less a foodstuff at the end—were hidden, like precious members of the family, from desperate marauders who raged through the city in search of food.

During the fight against the Spanish and their Indian allies, the rigid, complex communal structure of Aztec society began to break down. The domi-

Cortés'arrival in Mexico. Artists and mapmakers in Europe thought of the moment when Europeans first arrived in the New World as a scene of friendly greeting by largely naked natives. The reality was often far more cautious, and sometimes decidedly less friendly.

The conversion of the natives was an integral aim of French and Spanish colonial enterprise (above left)*. The deep gratitude of the converted was then taken as a justification for European conquest and domination. Here, St. Francis Xavier is shown baptizing an Indian; from a relief at University College, Grenada, Spain.*

John White, who was the governor of the Roanoke colony in 1577, painted this watercolour, (above right)*, of a Virginia native woman holding a basket of corn.*

nant warrior élite (*teteuctin*), hereditary nobility (*pipiltin*) and high priests, had exerted control over the commoners, who were themselves divided into *calpulli*, or clans or wards, which were responsible for the many administrative, economic and military duties in the community. The serfs, who composed a third of the Aztec population, and labourers, who were not members of *calpulli*, had below them, at the very bottom, slaves, perhaps not more than 5 per cent of the total population. In the siege which finally destroyed the Aztec empire, one sees the "roofless" homes and once proud and busy roads lie like "broken shafts". With their communal structures in ruins, and abandoned by their gods, the Aztec, demoralized and silent, faced the new world created by the conquistadores. Even mild European diseases proved lethal to the Indians. Along with the excesses of *encomienda* labour, illness caused a drastic depopulation in the early decades of Spanish rule.

Thousands of miles to the north, a pastor in the Puritan settlement at Malden, Massachusetts, invoked a world of sorrow and lament:

" Mean men lament, great men do rent
their robes and tear their hair;
They do not spare their flesh to tear
through horrible despair.
All kindreds wail; their hearts do fail;
horror the world doth fill
With weeping eyes and loud outcries,
yet knows (or know) not how to kill."
(Michael Wigglesworth, *The Day of Doom*, 1662)

The arrival of the English in Virginia, in a map which records Indian fishing techniques, a palisaded settlement, sunken sailing ships and a sea monster. The mixture of imagined scenes and sharp observation makes these early maps a window into the European mind in the period of exploration and early colonization.

Wigglesworth sought to represent in simple, vivid language the day of Christian reckoning, when the atheist, bold and doubtful, felt the cold steel of damnation. It was a moment when all men were reduced to terror and dismay. Dignified civic leaders screamed and tore at their flesh in "horrible despair". The New World, so often described by European promoters as a bountiful garden, a glorious Eden, was a continent of tears.

The destruction of the indigenous cultures which followed the intervention of Europeans into their world proceeded as though by plan. Yet, such planning and thoughts of "final solutions" belong to the 20th century, not to the 16th. The war dogs of the Spanish, their terrifying horses, cannons, steel swords and native allies skilled in betrayal and ambush, proved invincible in Mexico. The Spanish intended to conquer, and grow irresistibly rich in the process. Unlike the Arabic and European slave traders in Africa, they did

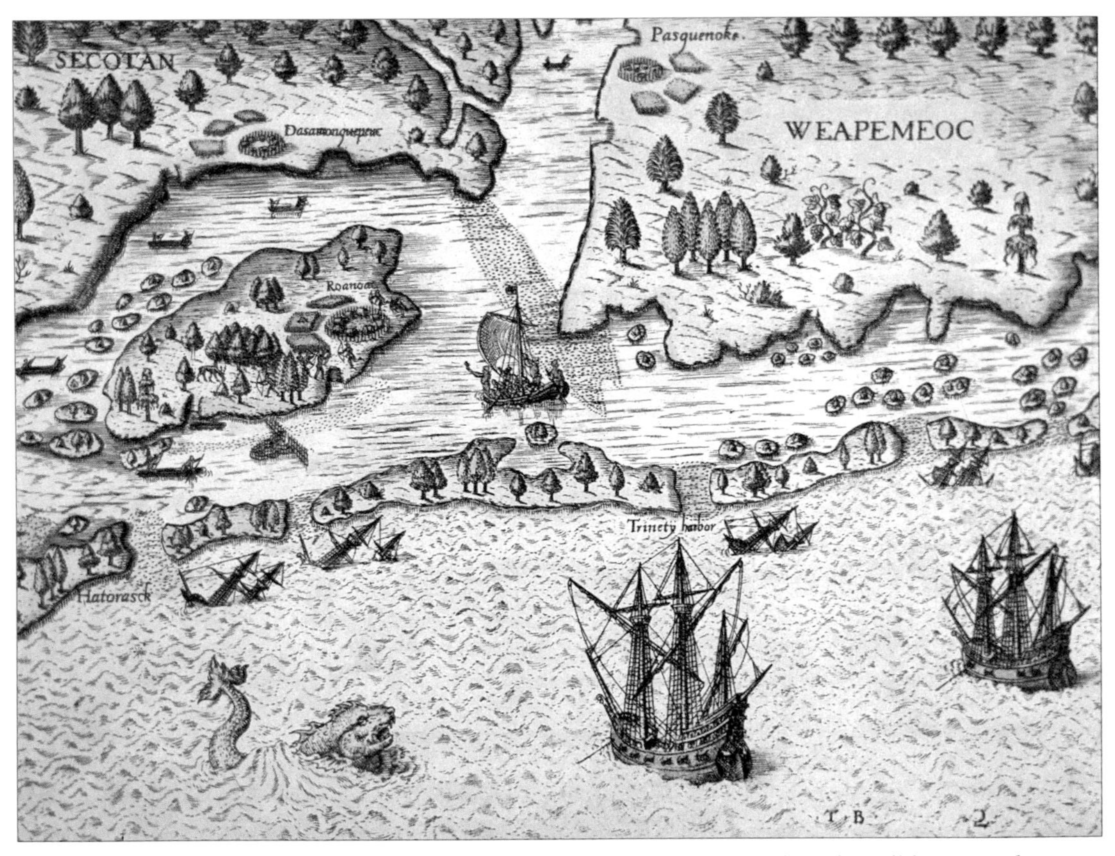

not seek to enslave the native population, but they did expect the conquered to labour for their masters, comforting themselves (if the thought ever arose) that though the masters had changed, the system of tribute had not. No other life was imagined for the natives. The system of rule which the Spanish created in Mexico and the Caribbean was one of forced labour.

The Dutch, French and English established no such rule over the wild territory they acquired to the north. The level of native civilization had nowhere reached that of the sophisticated Aztecs. There was no system of social control and rigid hierarchy which they could turn to their own purposes. The tribes they encountered were small, autonomous groups, strongly differentiated by language, and who possessed a lively sense of competition and rivalry with their neighbours.

The French found the Montagnais, Iroquois and Huron Indians fierce, brutal and scornfully indifferent to the piety of the Jesuit "black robes" who were sent to convert them. The French fleets that visited the inshore fishing banks during the summer months only came into brief contact with the Indians when they came ashore to dry and salt their catches. When traders began to penetrate the Gulf of St. Lawrence to acquire the fur pelts so cherished across Europe, their activities demanded only the most rudimentary of contacts with the natives. The fur trade soon drew the French a thousand miles into the interior, but they were only there on sufferance because they had goods to trade with the Indians.

The massive Spanish occupation and exploitation of Mexico, and the purely commercial and superficial French presence in the St. Lawrence, were alternative forms of the contact made by Europeans with the indigenous peoples of North America.

This watercolour by John White shows a Virginia Indian chief in decorative body paint.

Throughout the colonial period, which did not end in Mexico until the 1820s, the Spanish in Mexico were largely without direct rivals (Spanish Caribbean possessions and coastal settlements were repeatedly sacked by the British and French). But the great distances which separated New Spain from the nearest British colonies effectively removed them from the poisonous rivalries of the struggle between the French and the English for dominance in North America.

The early history of the colonial settlements of Britain and the Netherlands is marked by periods of more or less peaceful trade with the nearby natives which sometimes broke down over disputes over land. Minor irritations had a stunning capacity to end in violence. The sale of brandy to the natives was repeatedly banned, but continued scarcely without interruption. Reprisals followed, and then outright massacres. The Pequot tribe, who had driven the Niantics out of the area around New London, Connecticut, were attacked and dispersed by English colonists in 1637, effectively assuring peace until the 1670s.

The Dutch were soon drawn into conflict with the Indians around them. "On the east side, upon the main land", wrote Johan de Laet in 1625, "dwell the Manatthans, a bad race of savages, who have always been very obstinate and unfriendly towards our countrymen. On the west side are the Sanhikans, who are the deadly enemies of the Manatthans, and a much better people... ." After unsuccessfully demanding the surrender of an Indian accused of murder, the Dutch made a night attack on refugee Indians at Pavonia in 1643. They were not as successful as the English in eliminating the danger, and were soon attacked by all 11 tribes in the region, which left Dutch farms in ruins and the 500 colonists cowering within their palisaded fort.

The growing British population, and their insatiable hunger for land, brought them into direct conflict with the Dutch who occupied the territory between Pennsylvania and Massachusetts. With a fort at the southern tip of Manhattan and Fort Orange at the head of the Mohawk Valley, the Dutch controlled the finest navigable access to the interior of the continent. The Dutch position on the Hudson challenged the French for control of the fur trade, and they occupied land which the English colonists regarded as theirs by right.

The Director General and Council of New Netherland were faced with the encroachments by English settlers on territory claimed by the Dutch at De Hoop (New Hope, later Hartford, Connecticut) and on Long Island. In May 1640, a party of 25 soldiers was sent from New Amsterdam to North Hempstead on Long Island, where they arrested six men. The records of the interrogation of these Englishmen, born in "Bockingamshier" (Buckinghamshire) and "Lingconschier" (Lincolnshire) contained the only answer to which the Dutch had no response:

> "What did they [the English] propose doing there, and how many people were to come there?"
> "They intend to plant, and if the place was good, a great many more were to come."

A year later the Director General sent 50 soldiers in sloops to fortify their position on the Connecticut River. Expostulations, polite letters of protest, trade bans, and armed reassertions of the Dutch title to the Connecticut territory made little difference. In 1664, King Charles II issued letters of patent in which he granted the entire region from Connecticut to Delaware, including the territory of New Netherland, to his brother James, Duke of York. (New Amsterdam was renamed in his honour.) Four English men-of-war arrived off Staten Island to enforce the claim. The Dutch could not resist, and in the peace treaty which followed they ceded their interests in

Christoph Weiditz's Indian figures placed American natives into culturally familiar contexts for Europeans. The tamed wild parrot suggested that the natives inhabited an unspoiled Eden.

The Puritans were well-armed, militant and well-organized. The procession to Sunday worship, (above), *has about it a mixture of piety and vigilance. Relations with Indians improved, but the Puritans remained cautious of the threats potentially lurking in the dark New England woods.*

North America for the sugar-rich colony of Surinam.

After the swallowing of New Netherland, the territory of the British ran from the undefined region above Massachusetts to the edge of Georgia Colony, settled in 1733. It was a populous colonial realm, avidly reaching out into the wilderness for new land. Ships crowded the ports of Boston, New York and Philadelphia. Where the British came into contact with the French, whose trade routes had taken them beyond the Great Lakes, the seeds were sown for the Seven Years' War that began in 1755 and left the British triumphant across the whole of the inhabited territory north of New Spain.

"Let England knowe our willingnesse
For that our work is good;
Wee hope to plant a nation
Where none before hath stood."
Popular verse, London, 1610.

Charles II (1630-85), king of England from 1660, had an immense impact upon the development of the British colonies in North America. The portrait (right) *appears on the charter granted to Hudson Bay Company in 1670. Charles II made sweeping assignments of land (5 million acres between the Rappahannock and the Potomac to Lord Hopton in 1649; the whole of the Dutch territory in (present) New York to his brother James, Duke of York, in 1664), and granted many charters for colonial development.*

Cortés and the Conquest of Mexico

The Spanish victory in Tenochtitlán ended the greatest of all the Mesoamerican empires.

This is perhaps the most authentic portrait of Cortés, drawn by Christoph Weiditz who encountered the conquistador in Spain in later life.

An expedition sent by the Spanish governor of Cuba in 1518 to the Gulf Coast of Mexico returned with news of a great and rich kingdom inland. Hernan Cortés was commissioned by his brother-in-law, Governor Velásquez, to lead an expedition to the kingdom of the Aztecs. Cortés put up two-thirds of the risk capital for the venture, and recruited men—veterans of earlier expeditions—who were willing to share the risk. Five hundred and eight soldiers, armed with Toledo swords, crossbows and muskets, and accompanied by 16 horses and 100 seamen, anchored off San Juan de Ulúa in April, 1519. On August 16, Cortés set off inland for Tenochtitlán, the capital of the Aztec empire.

Cortés' men had advantages, like the horse, which compensated for their small numbers. But their greatest advantage lay in the leadership of Cortés and in his remarkable grasp of the nature of the struggle in which they were engaged. Cortés sought to terrorize the Aztecs and, by deft diplomacy, demoralize their will to fight. By recruiting disaffected tribes, like the Tlaxcalans—who had long harboured hatred towards their overbearing neighbours—he found the "geological fault" which enabled his band to overturn a great civlization.

Welcomed at Tenochtitlán, Cortés made the Aztec *tlatoani*, Montezuma, his prisoner. Disabling the structure of leadership, overturning the Aztec idols, forbidding human sacrifice, and the installation of Catholic symbols in place of the Aztec gods, gave Cortés the initiative throughout his campaign. Forced to retreat from Tenochtitlán after an Aztec uprising on June 30, 1520, Cortés led a brilliant campaign to surround and besiege the city. The 75-day siege did not end until the great capital, on orders from Cortés, was destroyed. The victory of 13 August, 1521, was the decisive moment in the long history of the Mesoamerican peoples, the moment when their introduction to European values, capitalism and Catholicism began.

Tlaxcala Indians greeting Cortés with a symbol of peace. The drawings, from Diego Duran, Historia de las Indias de Nueva Espanã*(first published 1880), were made by Mexican artists between 1560 and 1580. Spanish troops soon found that European body armour was too hot and cumbersome for the Mexican climate. Holding Cortés' horse is an African member of the company, perhaps the leader's servant and slave.*

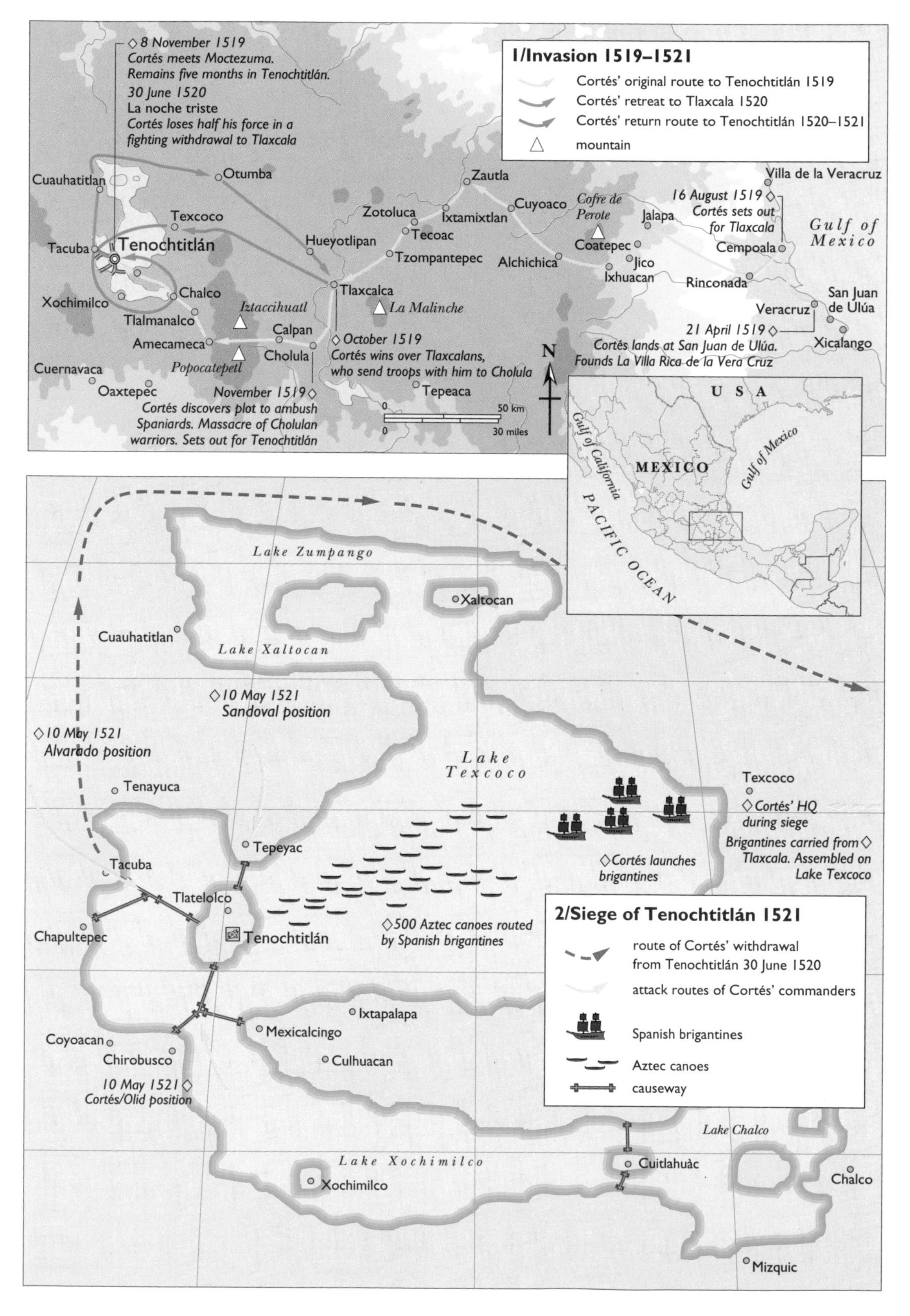
1/Invasion 1519–1521
Cortés' original route to Tenochtitlán 1519
Cortés' retreat to Tlaxcala 1520
Cortés' return route to Tenochtitlán 1520–1521
mountain
8 November 1519
Cortés meets Moctezuma.
Remains five months in Tenochtitlán.
30 June 1520
La noche triste
Cortés loses half his force in a fighting withdrawal to Tlaxcala
Cuauhatitlan
Otumba
Texcoco
Tacuba
Tenochtitlán
Xochimilco
Chalco
Tlalmanalco
Iztaccihuatl
Amecameca
Calpan
Cholula
Popocatepetl
Cuernavaca
Oaxtepec
November 1519
Cortés discovers plot to ambush Spaniards. Massacre of Cholulan warriors. Sets out for Tenochtitlán
Hueyotlipan
Zotoluca
Tecoac
Ixtamixtlan
Tzompantepec
Zautla
Cuyoaco
Tlaxcalca
La Malinche
October 1519
Cortés wins over Tlaxcalans, who send troops with him to Cholula
Tepeaca
Alchichica
Cofre de Perote
Coatepec
Jico
Ixhuacan
Jalapa
16 August 1519
Cortés sets out for Tlaxcala
Villa de la Veracruz
Cempoala
Rinconada
Veracruz
San Juan de Ulúa
21 April 1519
Cortés lands at San Juan de Ulúa. Founds La Villa Rica de la Vera Cruz
Xicalango
Gulf of Mexico
N
0 50 km
0 30 miles
USA
MEXICO
Gulf of California
Gulf of Mexico
PACIFIC OCEAN
2/Siege of Tenochtitlán 1521
route of Cortés' withdrawal from Tenochtitlán 30 June 1520
attack routes of Cortés' commanders
Spanish brigantines
Aztec canoes
causeway
Lake Zumpango
Xaltocan
Cuauhatitlan
Lake Xaltocan
10 May 1521
Sandoval position
10 May 1521
Alvarado position
Tenayuca
Lake Texcoco
Texcoco
Cortés' HQ during siege
Brigantines carried from Tlaxcala. Assembled on Lake Texcoco
Cortés launches brigantines
Tepeyac
Tacuba
Tlatelolco
Chapultepec
Tenochtitlán
500 Aztec canoes routed by Spanish brigantines
Ixtapalapa
Mexicalcingo
Coyoacan
Chirobusco
Culhuacan
10 May 1521
Cortés/Olid position
Lake Chalco
Lake Xochimilco
Cuitlahuàc
Xochimilco
Chalco
Mizquic

Spanish Empire in the West

After the fall of the Aztecs, Spain ruled a vast empire in the New World which was to last for centuries.

"*It is not Christianity that leads them on, but rather gold and greed.*" Said by the devil in Lope de Vega's play *The New World*

After the conquest came the spoils: a great empire containing 25 million Indians, and a land rich in promise. The Spanish crown had given considerable latitude to the conquistadores, and acquiesced in the extension to Mexico of the *encomienda*, the award to the faithful warriors of Indian labour in return for the care of the Indians' souls and well-being. Indians were not chattel slaves, but the carefully organized exploitation of the *encomienda* system set the pattern for the systematic exploitation of the Indians by the Spanish.

Relations were literally murderous: contact with Europeans brought to a people isolated from the rest of the world virulent diseases such as smallpox, measles, influenza and plague. The result was a catastrophic depopulation of the new Spanish domain. From 1519 to 1532, the Indian population fell from 25 million to under 17 million. Over the next 16 years, to 1548, it fell to 6.3 million; in 1568 to 2.65 million; and in 1580 to below 2 million.

The Spanish built cities, constructed according to precise instructions from Madrid, as administrative centres of their rule. They sent out explorers to inspect the land to the north, who in time reached Santa Fe and Miami. The Catholic Church built convents, churches and missions. The crown sought to reassert its power over the "men of the sword" by the imperial bureaucracy—the "men of the pen". The day of the conquistador had ended, and the long period of meticulous Hispanic bureaucratic control of the land stretching from Panama almost to the banks of the Mississippi had begun.

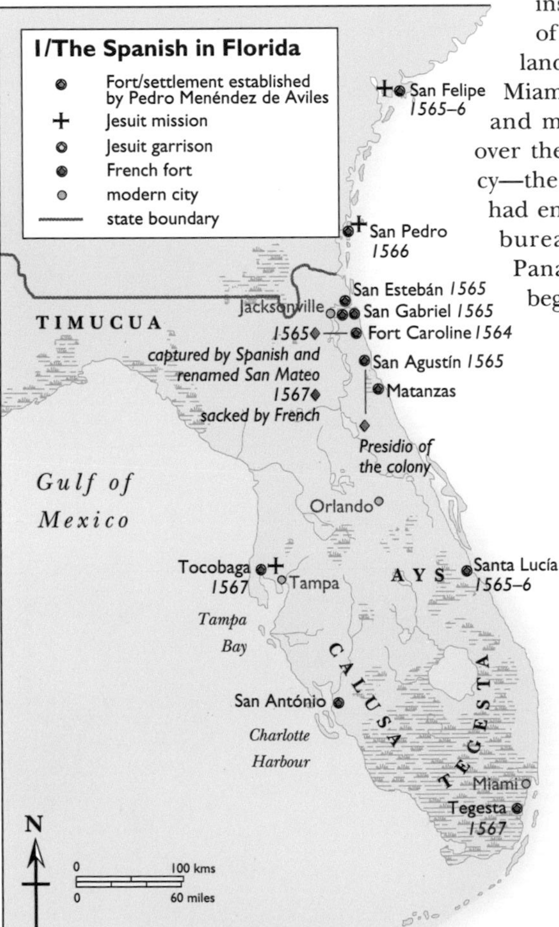

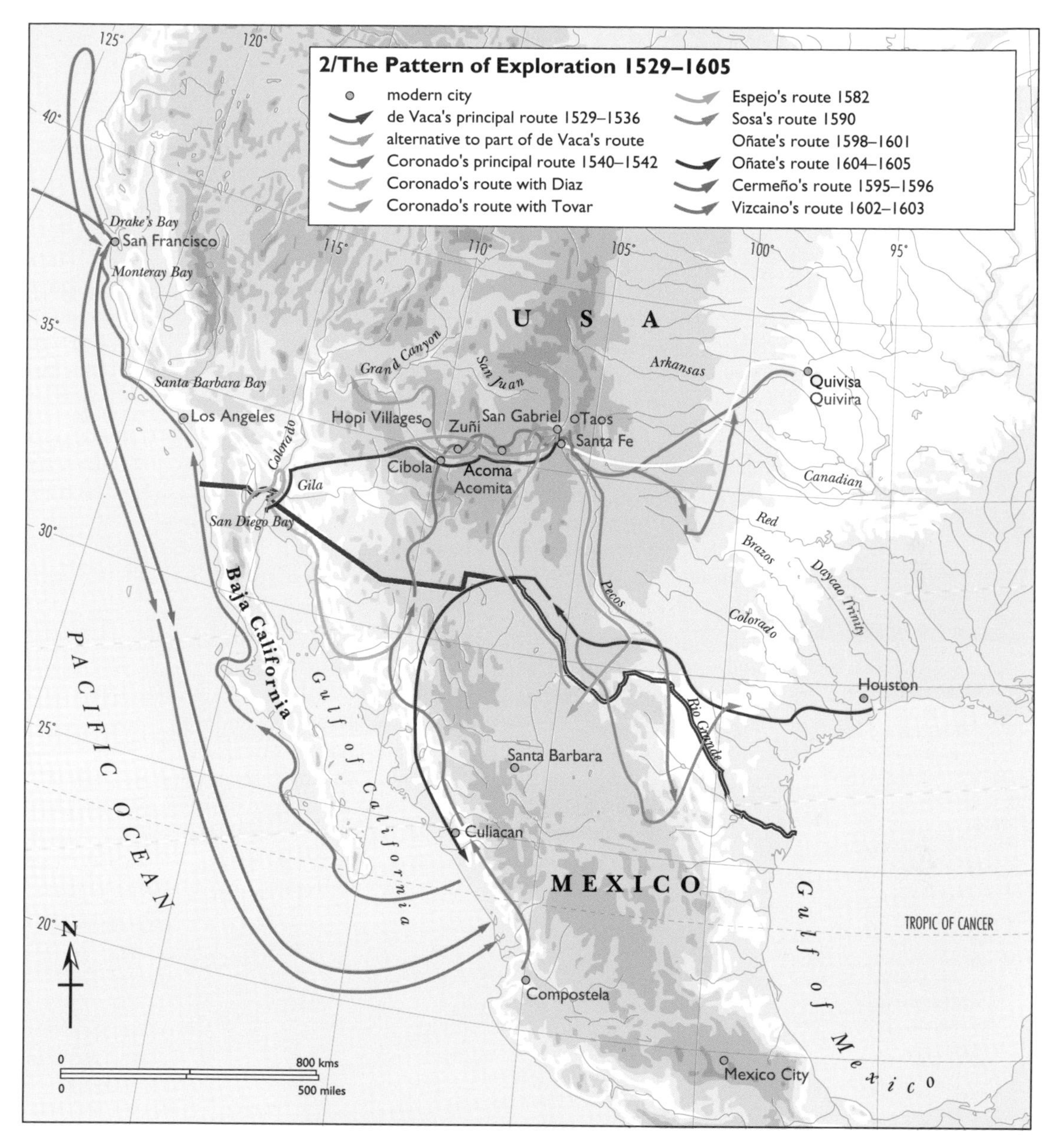

Right: *Cortés adopted the Aztec system of tribute and distributed Mexican towns as rewards to his soldiers. The* encomienda *system was a gift of care in exchange for the labour of the natives. The requirement of tribute (recorded here by native picture-writing) reduced the natives to serfdom.*

Left: *After the fall of Tenochtitlán in 1521, the city was renamed Mexico City and rebuilt by Spanish artisans in accordance with imperial edict.*

French Empire of the St Lawrence

France sought a new model for colonization, but was unable to withstand the rapidly-growing power of Britain in North America.

The voyageur, from a French engraving made in 1722. Despite the absence of snow, he is wearing snowshoes. They travelled deep into the northern forests to trade for pelts.

The French came to Canada around 1500 to dry and cure the cod that their fishermen caught in the inshore fishing grounds. Seasonal visits were merely a convenience, and left little mark upon the land. The offshore fishermen returned directly to France with "wet" (i.e. salted) cod. Trade in furs drew French traders into the Gulf of St. Lawrence. Jacques Cartier attempted a permanent settlement near Québec in the 1530s, but harsh conditions and hostile natives caused the colony to be abandoned. The Spanish model of conquest, settlement and exploitation was irrelevant to the conditions of New France. A new model was needed.

Samuel de Champlain established Québec in 1608 as a base for trade with the Huron and Algonquin tribes. By 1685, the French had established trading forts on the Mississippi. Their efforts at colonization were at best half-hearted. The trade in furs and cod did not require the development of the economy. Settlers in Acadia (Nova Scotia) and along the St. Lawrence were subsistence farmers, thinly spread across a vast territory.

The numerical superiority of the English colonies on the Atlantic threatened the French position, and military conflict led to the Treaty of Utrecht (1713) which awarded Acadia and Newfoundland to the British crown. The French sought to strengthen their position by the founding of New Orleans in 1718, with a plantation economy worked by African slaves. With a total population of less than 80,000 in the 1750s, they were unable to resist the British onslaught which led to the fall of Louisbourg, Québec and Montréal. In the Treaty of Paris of 1763 they realistically abandoned the entire territory east of the Mississippi. For the first time the British were unchallenged on the eastern half of the continent.

Cartier's landing in 1542 was drawn on this map, the Carte de Vallard, *made four years later. Fur-trimmed natives, the woodland hunt, and the great brown bears, captured the imagination of France.*

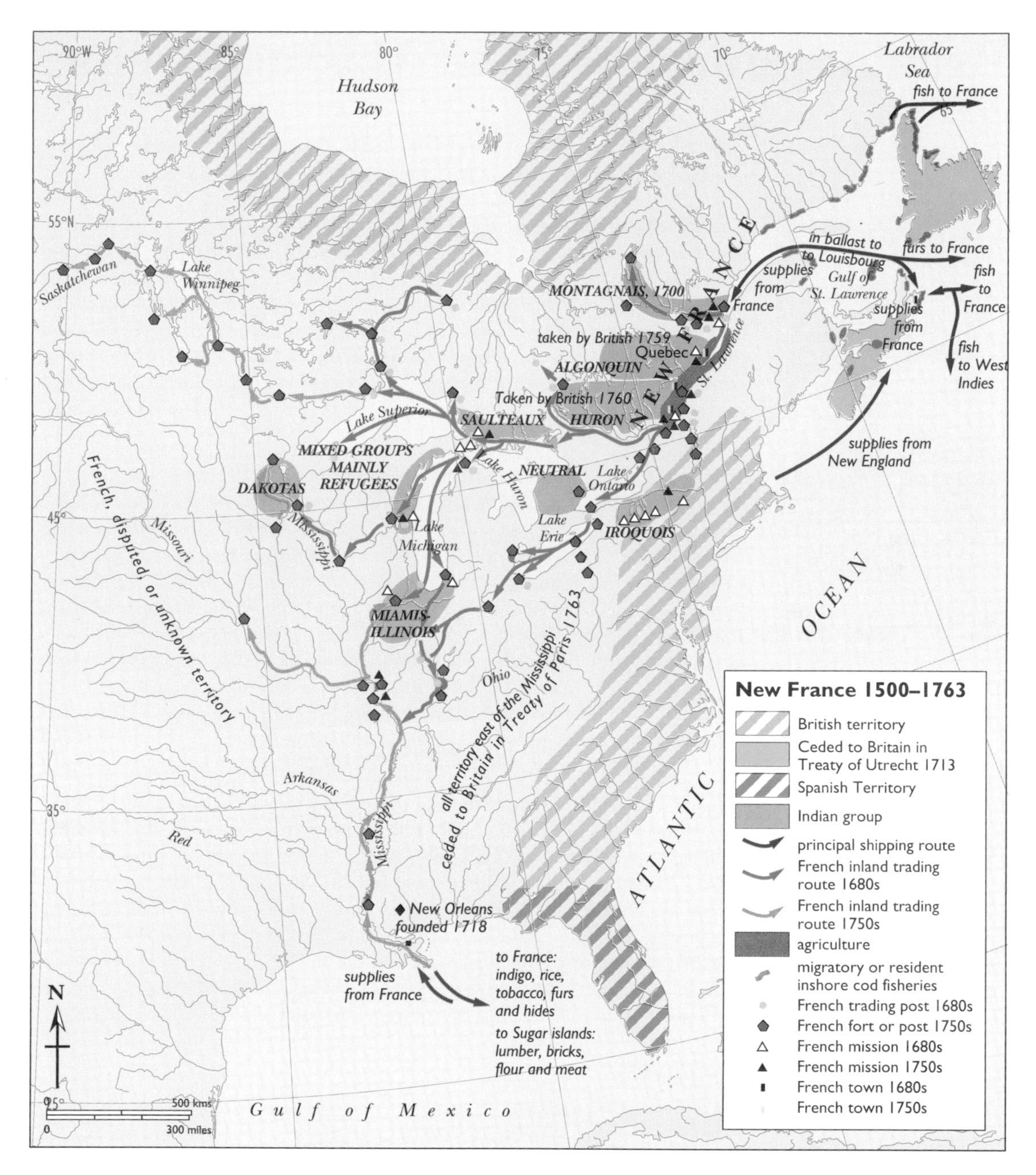

French maps in the 17th century became increasingly accurate and detailed, recording the location of Indian tribes and settlements (Hochelaga), as well as French missions and trading posts.

Britain: Jamestown and Massachusetts

The English colonies survived because they adapted to local conditions, and possessed a strong sense of shared purpose.

Colonial leisure: Virginia gentlemen enjoy a pipe and a glass, while slaves harvest the tobacco crop. Label for Virginia Tobacco, London, c. 1700.

The European powers came to the New World for reasons which they all shared: the pursuit of money, power, and the extension of national grandeur, but they approached the practical business of colonization quite differently. The Spanish came, conquered, converted and proceeded to settle and transform their colonies. The French, on the other hand, followed their fishermen and fur traders, establishing fortified bases from the St. Lawrence to the Mississippi, and made only half-hearted efforts to develop a populous community. Proselytism among the proudly independent northern tribes came to grief.

The English encountered nothing so overweening as the Aztec empire. The Algonquin tribes living along the Atlantic coast could be appeased, traded with, and, when necessary, routed and dispossessed. Like the Dutch on the Hudson Valley, they had little hope of converting the Indians to Protestant Christianity. There was no British gift of *encomienda* to support gentlemen loiterers in the New World. After a disastrous first winter at Jamestown in 1606-1607, half the "planters" died from disease and starvation. Men were needed with practical skills, who could be persuaded to exert themselves. The sturdy surviving settlers could only be governed through consent; democracy was an early crop in the New World. The headright allotment of land, which encouraged new emigration, strengthened the attachment of those who came from England.

Occupying an Indian-held territory, the Plymouth settlers were conscious of the need for defence against Indian attacks. As the colony grew, farms and homesteads were dispersed in small outlying villages, with sizeable acreage devoted to common land for livestock. The Plymouth Colony was incorporated into the Confederation of New England in 1692.

The New England settlers, drawn from extreme Protestant sects, were ideally suited to endure in a harsh wilderness. They possessed a strong sense of social cohesiveness, and a belief in the divine purpose of their labours. Sustained support from London enabled the Puritan plantation to survive. The remoteness of the link with England, and the lightness of the controlling hand, effectively left the Puritans to settle their own destiny. Political liberties, democratic representation and limitless free land encouraged emigration. The colonies were poor ventures when considered on a strict financial basis. But the decision to encourage agriculture, and the fortuitous discovery that tobacco flourished in the Virginia soil, gave the British a firm basis for their colonies.

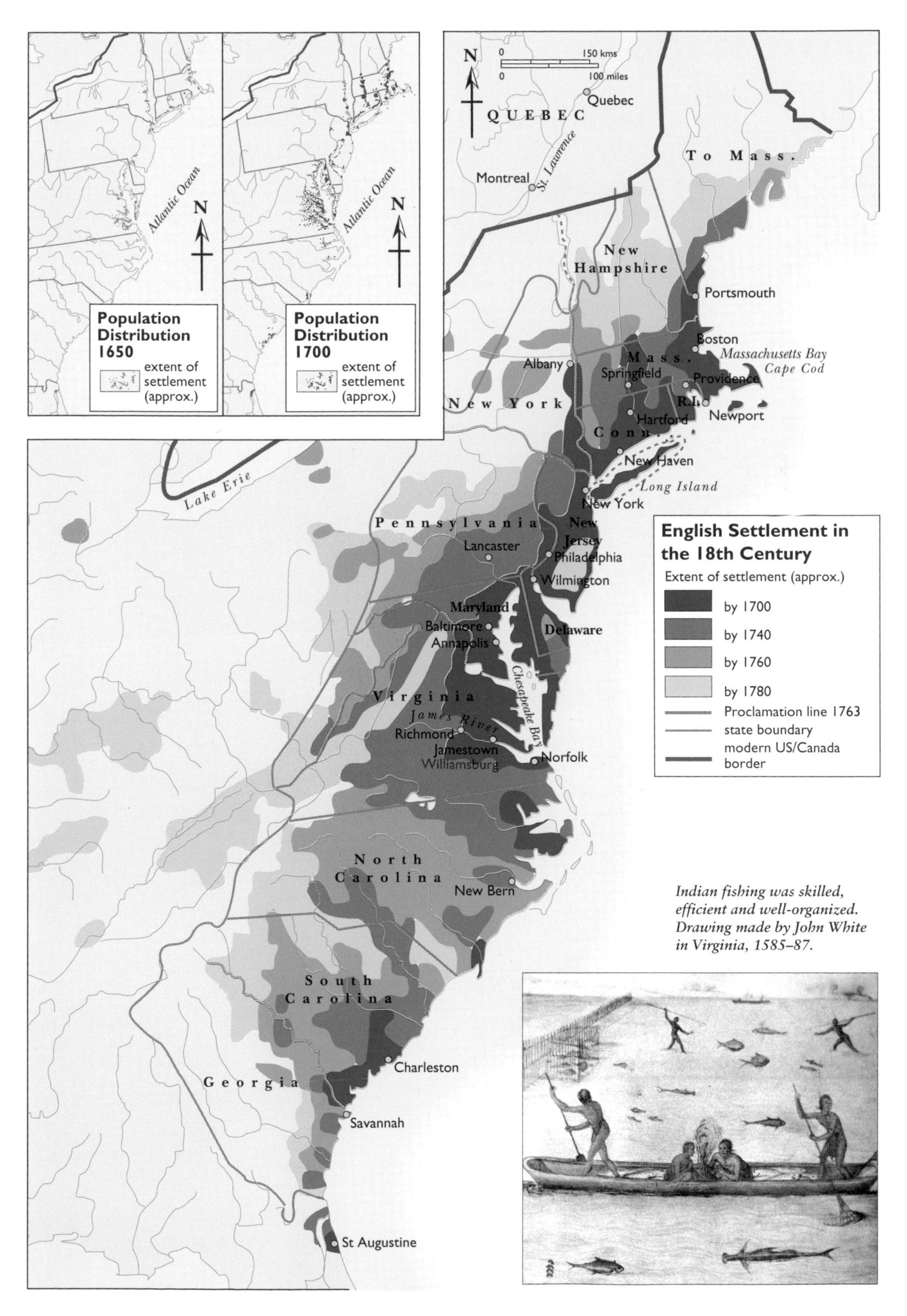

Indian fishing was skilled, efficient and well-organized. Drawing made by John White in Virginia, 1585–87.

Slave Trade

Slave labour existed in every colonial possession in the New World and in 1619 slaves were introduced to Virginia.

> *William Whittington sells to John Potts 'one Negro girl named Jowan; aged about Ten yeares and with her Issue...for their life tyme. And their Successors forever'.*
> Bill of Sale, Maryland, 1652.

Africans captured by Arab slavers or sold into bondage by tribal enemies had long formed a staple of commerce across the Mediterranean. In the 15th century, Europeans bought slaves in the great slave markets of Lisbon, Seville and Venice. In 1441 a Portuguese raiding party travelling along the African coast captured ten natives who had been sold into slavery, and returned them to Lisbon for sale; it was a lucrative cargo.

The Portuguese sent African slaves to their newly-captured possessions of Cape Verde, Madeira and Sao Tomé, where they were worked to death in the sugar plantations. When they began the colonization of Brazil in 1530, there was a greatly expanded market for slaves. (For every slave sold to the United States from 1500 to 1860, over six were sold to Brazil). Disease and exploitation in the Spanish Caribbean possessions soon left the authorities

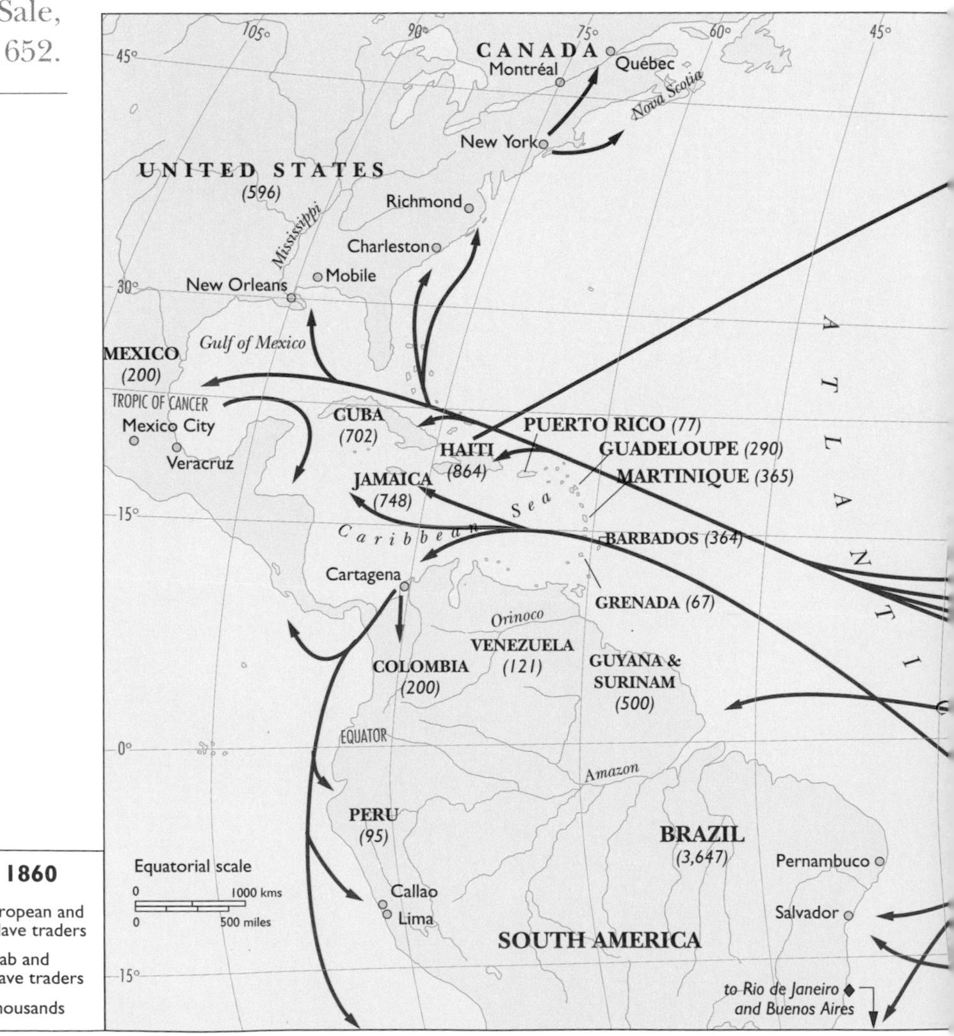

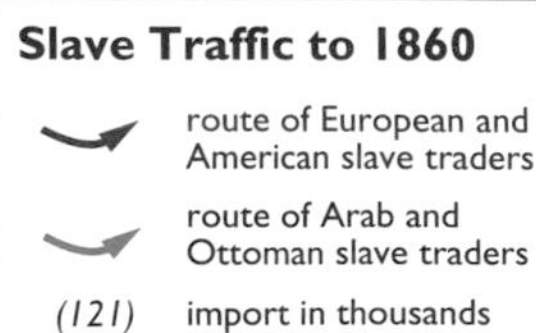

demanding slaves to perform work which had previouly been done by Indians. To prevent the harsh treatment of *encomienda* Indians in Mexico, Cortés allowed the importation of African slaves. In 1518 the Spanish king, Charles V, granted an *asiento* which confirmed the importation. From that date, the Atlantic slave trade began in earnest.

Dutch and English raiders, preying on Spanish and Portuguese galleons, soon challenged for control of the trade. In 1619, a Dutch privateer sold a cargo of slaves in Jamestown. Six years later, a similar cargo was landed in New Amsterdam (New York). When the Treaty of Utrecht of 1713 assured the English of a sale of 144,000 slaves per year to the Spanish colonies, they became the dominant slaving nation.

The tobacco traders at Jamestown, who had to rely upon indentured European labourers, welcomed slave labour. In 1662, the children of slaves were legally condemned to lifelong servitude. Even baptism could bring no escape from bondage. In the 18th century the legal restrictions which separated white Virginians from their slaves grew harsher and more systematic, planting the seeds for the full maturing of the slave plantation system after the War of Independence.

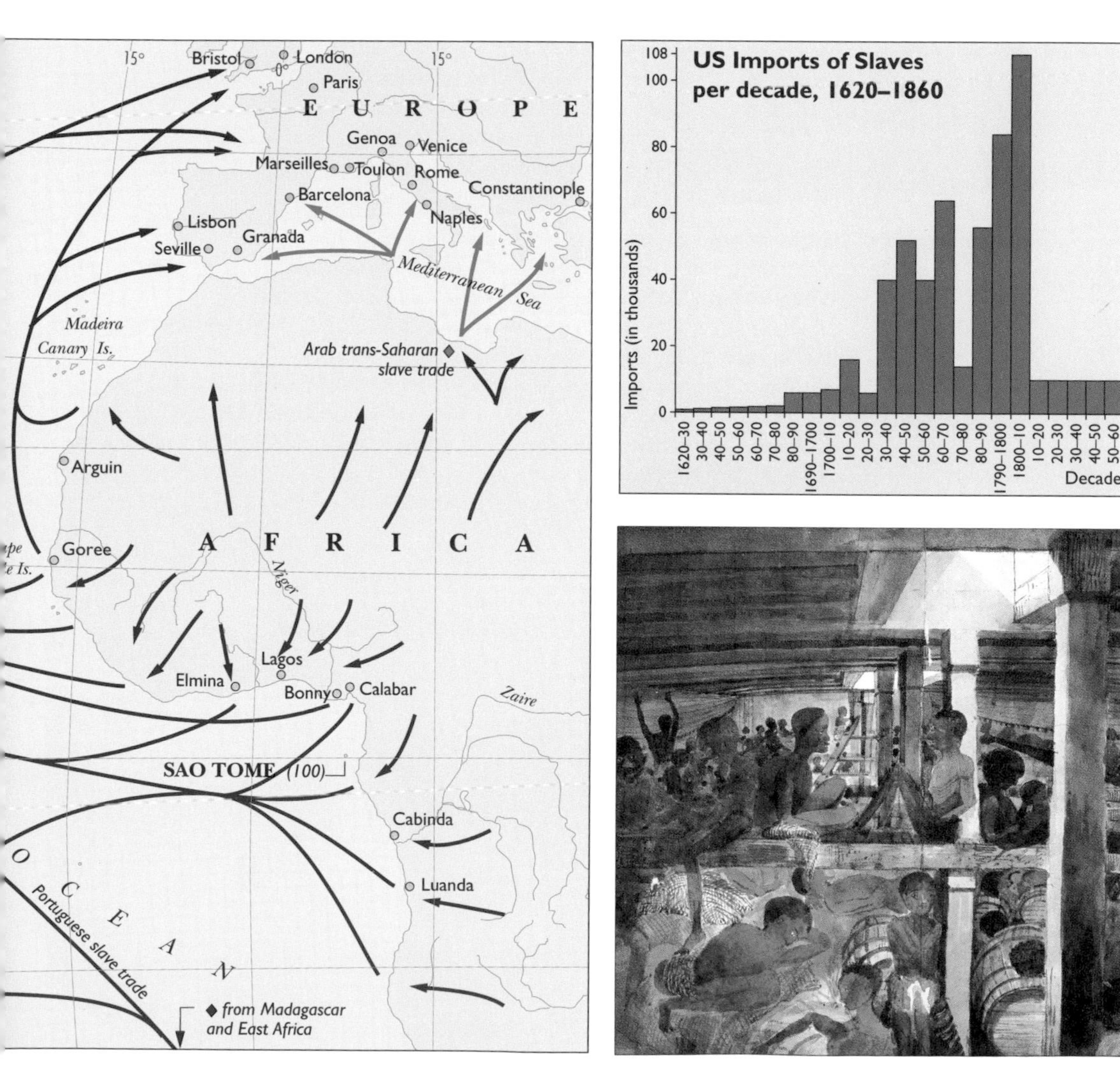

Below: *This watercolour was done by a young naval officer in a Spanish slave ship captured by HMS* Albaroz. *The atrocious conditions in the slave ships was among the causes for the widespread revulsion against the slave trade.*

Imperial Rivalries

A bloody battle in backwoods Pennsylvania began the Seven Years' War in 1755. At the end of the war, New France surrendered to the victorious British.

"The Marquiss [sic] de montcalm is at the head of a great number of bad soldiers. And I am at the head of a small number of good ones, that wish nothing so much as to fight him."
General James Wolfe, writing to his mother immediately before the Battle of the Plains of Abraham, 1759.

The rivalry between the British and the French extended across the planet, a chess game of bluff and counter-bluff, of bold assaults and cold-blooded massacre. The Indian allies on both sides in North America murdered civilians and tortured and sometimes killed captured enemy soldiers. Nothing less than control of the expanding world economic system was at stake. Together with their Indian allies in the 1750s, the French effectively blocked the advance westward of the British colonies which lined the Atlantic coast. With a total of 80,000 inhabitants in New France facing 1,250,000 British colonists, the struggle was doomed, although British military incompetence and brilliant French generalship did much to even the odds.

The Seven Years' War in Europe saw Britain subsidising the Prussians, who proved more than a match for the French, Austrians and Swedes. In the New World, the war began in July 1755 with an attempt by General Edward Braddock to dislodge the French from Fort Duquesne in western Pennsylvania. Braddock's death and the rout of his force of 1400 redcoats and 450 'blues' (provincial soldiers, including the young Major George Washington and the frontiersman Daniel Boone) began two years of military disasters.

Under Montcalm's fine generalship, the French defended Fort Duquesne and Crown Point, destroyed Fort Oswego and forced the surrender of Fort William Henry. When Pitt became British Prime Minister in 1757, the tide of war turned. Redcoats captured Louisbourg and Fort Frontenac. On 13 September 1759, with two devastating musket volleys on the Plains of Abraham, General Wolfe captured Québec. After Montréal was surrounded, the French position in North America became untenable. The French dream of an empire from the St. Lawrence to the Mississippi ended with peace in 1763. Acquiring East and West Florida from the Spanish, Britain was now without serious rivals in the New World.

Wolfe (1727–59) commanded the British forces which made a surprise landing to the north of Québec and defeated Montcalm and the French on the Plains of Abraham. Both generals died in the battle.

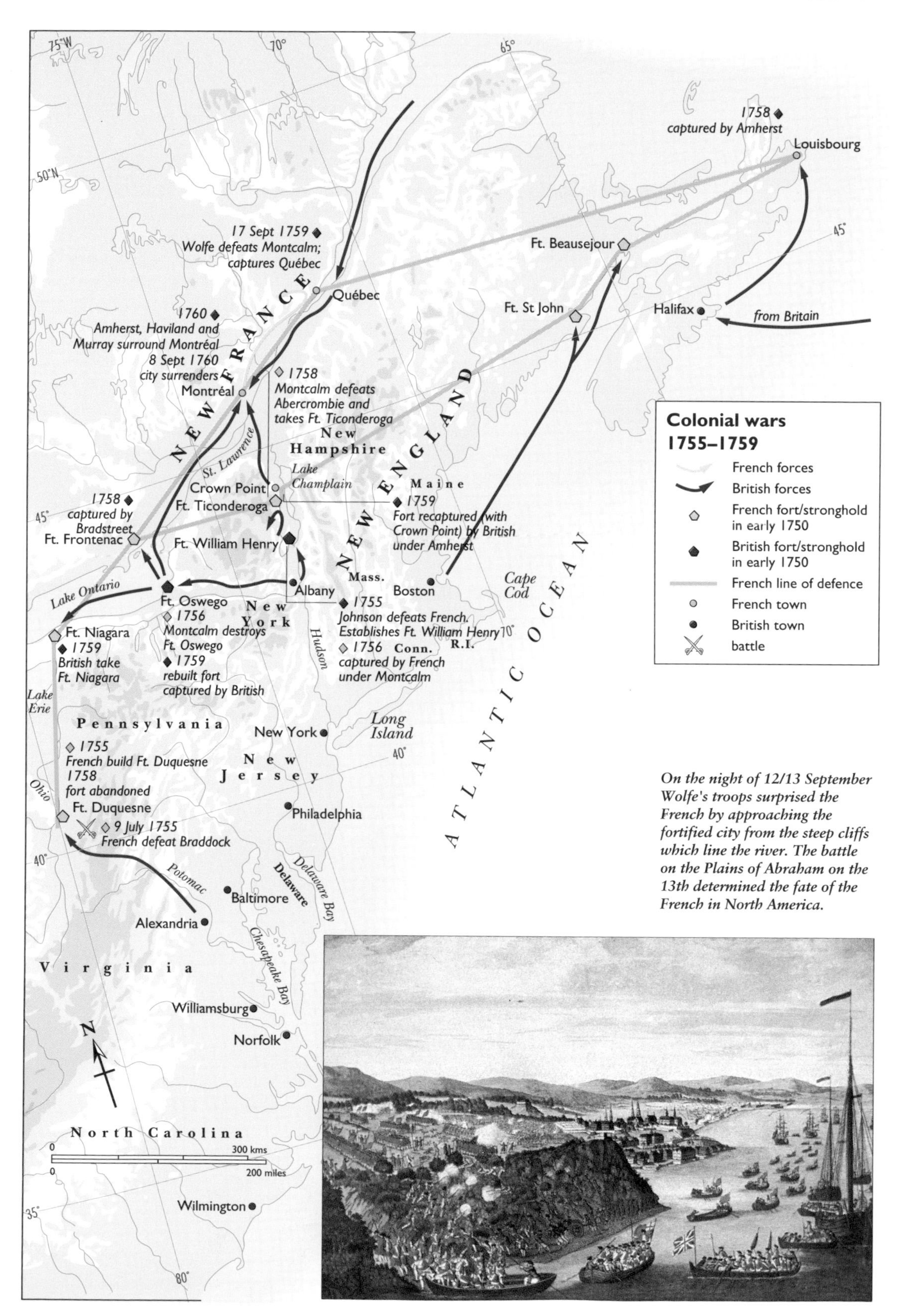

On the night of 12/13 September Wolfe's troops surprised the French by approaching the fortified city from the steep cliffs which line the river. The battle on the Plains of Abraham on the 13th determined the fate of the French in North America.

Immigrants

The first immigrants came as employees of trading companies, or as members of groups planning to settle together.

"From May to September 1607 ... fifty in this time we buried ...."
Captain John Smith, *The General Historie of Virginia,* 1624

Immigrants to the New World in the colonial era seldom came as individuals. They were employees of chartered joint-stock trading companies, and worked in occupations dictated by the company. They bought supplies from the company store. Furs and agricultural produce were sold to the company at fixed prices. There was little incentive for the settlers to improve conditions on their own. It was only when land tenure passed from company to individual ownership that the settlers possessed a real incentive.

Land was abundant, but could be improved by the exhausting labour of clearing forests. Unlike England, land was available at giveaway prices. With land so cheap, all of the colonial settlements were highly restless. Attempts in Massachusetts to recreate the familiar European nuclear village, surrounded by fields farmed in strips, soon broke down. There was so much land available that community organization and social control, even when the settlers were of a homogeneous composition, was hard to maintain. Groups repeatedly quarrelled with the leadership of the colony, and were exiled, or migrated voluntarily to form a settlement elsewhere.

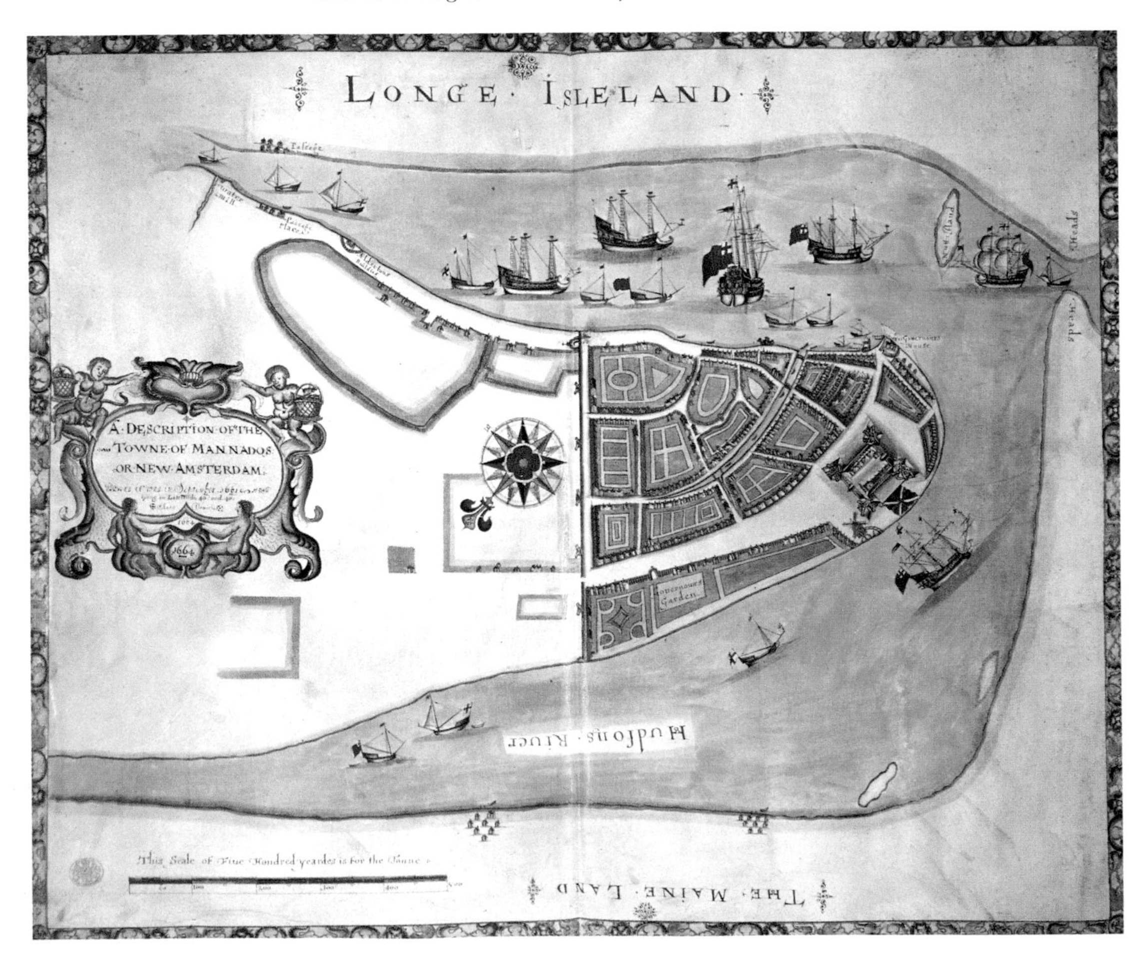

Above: *The dividing line between the commercial and residential property scarcely existed in early New York. The ground floor of this Dutch-era corner building constructed out of wood, with its canvas awning, and brick sidewalk, served as warehouse, small workshop and retail premises, while the proprietor's family lived above.*

Left: *The fur trade brought the Dutch to the Hudson River, and was their primary purpose in establishing the colony. New Amsterdam (map made 1664, the year the British took over the colony), located on the southern tip of Manhattan island, remained the largest Dutch settlement, and the administrative centre of the colony. The harbour on the East River encouraged development along the shoreline. The Hudson River side was lined by the Governor's Garden and farms. The northern limit of the settlement was marked by a wall (along the line of Wall Street today) to defend the settlement against native attack, but it never fulfilled that role and was soon allowed to decay.*

Among the immigrants were Jews expelled from Spain and Portugal in the 1490s, and Huguenots driven out of France in 1685, who crossed the length of Europe in search of a place of permanent settlement. Huguenots were welcomed by a Protestant prince in the Palatine in southwest Germany, and in Britain, before coming to New Paltz and New Rochelle, New York, from the 1670s. Self-contained groups, like the Dutch around Albany and the Huguenots at New Paltz, retained their own language for generations.

Jewish exiles found a haven in Holland, and when the Dutch seized coastal ports in Brazil in 1633, Jewish traders settled at Pernambuco and Recife. When the tide of war eventually changed, the Dutch were expelled and in 1654 a small group of Jews fled to the Dutch port of Curaçao. They subsequently arrived as penniless refugees in New Amsterdam. This was the beginning of the Jewish community in New York City. Due to the arrival of many immigrants from a variety of places in Europe, New York was a more ethnically diverse community than other American cities.

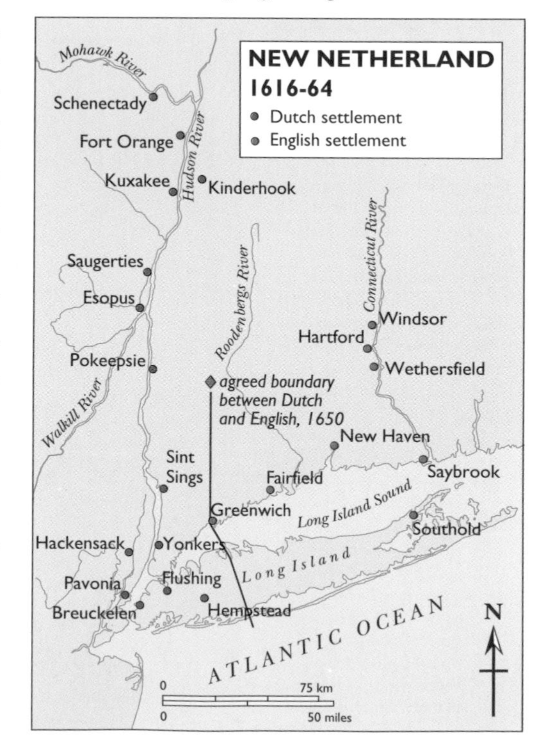

III: Making a Nation

In scarcely more than a generation the British colonial world, so triumphant after the defeat of France in North America, was turned upside down.

"We may please ourselves with the prospect of free and popular governments, but there is great Danger that these Governments will not make Us happy. God grant that they may!"

John Adams to James Warren, April 1776

A peace signed in Paris

The treaty which ended the Seven Years War was signed in Paris in 1763, four years after the heroic death of General Wolfe on the Plains of Abraham. Popular engravings of the death of Wolfe had gone on sale in London and throughout the colonies. Before the death of Horatio Nelson in 1805, Wolfe was the British Empire's greatest military hero and most poignant martyr. It was when his death was chosen by the American painter, Benjamin West, for a canvas on a monumental scale that the triumph of the British in North America found its greatest representation.

West's interest in such a contemporary subject was laden with difficulties. Sir Joshua Reynolds feared that the "inherent dignity" of the subject would be marred by the choice of "the modern garb of war". West's patron, George III, declined to purchase the painting, remarking that it "was very ridiculous to exhibit heroes in coats, breeches, and cock'd hats." The correct models for such a painting were classical; heroes wore togas, struck bold poses with sword and shield and did not die with Indians pensively kneeling in the foreground. The appropriateness of the representation, if not its literal accuracy, was open to public confirmation. Six individuals who appear in West's painting were alive and in London when it was first exhibited at the Royal Academy in London in 1771 (four requested West to make copies for them). Britain, however, was not yet ready for the heroic repre-

By March 1770 the soldiers in the British garrison in Boston and bands of civilians were both looking for trouble. There was no sign, even after the Townshend duties were altered and the Quartering Act was allowed to lapse, of any serious determination to resolve the political crisis. The mood in the city became increasingly ill-tempered and when a beleaguered sentry at State House summoned soldiers to disperse an angry crowd, they fired upon the Bostonians, killing 3 outright. Two died later of wounds. A general uprising was avoided only when the Governor ordered the British garrison to withdraw from the city. The commanding officer of the troops and four soldiers were tried for murder and acquitted. Paul Revere, a patriot silversmith, created an image of the "Massacre" (right) which outraged all patriots and brought the final breakdown of relations between Britain and the colonies appreciably closer.

After serving as aide-de-camp *to General Washington, John Trumbull (1756–1843) retired from the Continental Army in 1777 and went to London to study art. He was imprisoned for seven months as a rebel spy and then deported, but was able to return to London in 1784 where he worked on a painting of the* Battle of Bunker Hill *in the studio of Benjamin West. He also did a* Death of Montgomery *and a* Surrender of Cornwallis. *Trumbull's* Signing of the Declaration of Independence *was painted after his return to America in 1789. It is in the rotunda of the National Capital in Washington, D.C.*

sentation of scenes from contemporary life. It took an American painter, and his disciples among the painters of a younger generation, John Singleton Copley and John Trumbull, to confirm the contemporary as a fit subject for the painter. It was a small sign of the ways in which the colonial culture and that of the mother country were beginning to diverge.

At the end of the war diplomats settled down to share out the spoils. The French, forced to accept military defeat, ceded to Britain their claims to Acadia, Cape Breton, Canada, and the islands of the St. Lawrence River. All territory east of the Mississippi (with the exception of New Orleans) was given to Spain. In return, the French West Indies, conquered in 1762, were returned by the British. St. Vincent, Dominica and Tobago were restored to Britain. The Spanish re-acquired Cuba in exchange for their claims to East and West Florida.

There were perhaps as many as 2 million persons, free and slave, residing in the American colonies in 1763. The growth of urban centres was a sign of developing prosperity. Boston, with 20,000 inhabitants, was the largest Northern town. Charleston, with half that population, was the largest in the South. The Empire assured markets for American exports, but restrictions upon manufactures in the colonies—which were designed to protect against unemployment at home—served to restrict the extent and kind of industrial development permitted. The manufacture of wool, either raw or woven as cloth (1699), beaver hats (1732) and iron (1750) was sharply restricted or banned altogether. American ship builders and masters proved more than equal to the task of competing with the commercial fleets of Great Britain. The 20,000 gallons of rum produced annually in New England (there were 63 distilleries in Puritan Massachusetts Bay), as well as the wheat, flour, rice,

Like John Trumbull, Charles Willson Peale (1741–1827) had studied at the studio of Benjamin West in London. He became one of the most commercially successful painters in colonial America. It was while serving in the Continental Army that Peale found his great subject—portraits of military and political leaders. He completed 14 likenesses of George Washington, including the earliest known portrait—Peale's Washington at Princeton *(above). Scarcely a president was allowed to take the oath of office without sitting for Peale. His brother, and three of his 11 children—all named after artists—became celebrated painters.*

tobacco and furs exported through New York and Québec, made the colonial economy a prosperous one. Tobacco planters lived on spacious estates. The homes of sea captains, lawyers and merchants that lined Wall Street and Broadway in New York were large, handsome structures.

The Union Jack waved over a vast territory extending from Hudson Bay to Florida. Everywhere political élites alone enjoyed the right of franchise, and vied for control of public offices. The system of governance was not uniform. There remained corporate, proprietary and royal colonies, each with a slightly different role for the colonial governor, and each functioning within a different pattern of law and political development. Nonetheless, that colonies should be self-governing lay at the heart of the system over which the British presided. The colonials expected the "liberties of English men" to be respected.

The French population of Québec—"alien" only in the eyes of their British conquerors—posed a particularly difficult problem for they constituted a substantial majority of the population. The French legal system, the French language and Roman Catholicism effectively disenfranchised the French *Canadiens,* who were ineligible to vote or to sit in an elected assembly. If the French population were ever to be incorporated into the new state, the traditional local élites and the Catholic hierarchy would have to be accepted. The Québec Act of 1774 went part way towards this goal. No provision was made for an elected assembly, but the ancient boundaries of Québec were restored. Freedom of religious observance was established; property rights and the seigneurial system were preserved. But the forcible expulsion of French Acadians, and their remote dispersal across the continent, haunted British policy for many years.

While containing substantial "foreign" populations (Germans in Pennsylvania, the Dutch in New York, Spanish-speakers in Florida), the essentials of English civilization—in the words of Arthur Meier Schlesinger—united the widely-dispersed inhabitants. "They read English literary works, cherished English political values, followed English commercial practices, paralleled English religious beliefs, adopted English educational methods, imitated English architecture, sang English songs, copied English dress, played English games." Nonetheless, developments within this overwhelmingly Anglophone world brought British rule to crisis and to violent collapse within a generation. Nothing seemed less likely when the Treaty of Paris was celebrated in 1763.

What mighty force could disturb this prosperous and secure scene? The roots of the conflict were found not only in the changing colonial policy, financial needs and mercantile outlook in Britain, but also in the differences which increasingly separated the colonies from the mother country. The failure to recreate a landed aristocracy in the New World encouraged a more democratic, self-reliant and optimistic approach to social issues.

Positions of preferment could seldom be passed on to sons.

In the towns, colonial rule was highly visible. In New York the mayor, recorder, sheriff and common clerk of New York City all owed their appointments to the governor, and thus indirectly to the crown. They and their families formed something of an imperial garrison within the city—and an obvious target for discontent. The King's birthday and other royal occasions were opportunities for patriotic military displays designed to reassert the crown as the symbolic source of authority. Taxes were collected at the ports, and it was in the urban areas where discontent found readiest expression. The further people lived from the cities, the less attention need be paid to the ceremonial deference owed to the governor. Of the struggles between factions in Westminster, nothing was known, nothing cared. Only those in hope of preferment attended to such matters. Once the threat of French-inspired Indian attacks had ended, for the farmer on the fringes of settlement the colonial mentality became attenuated, the old loyalties more theoretical.

British mercantilist policy sought, in ever-increasing detail, to restrict colonial commerce. While these regulations had been in force since the beginning of British colonial settlement in North America, the most energetic entrepreneurs inevitably came to regard colonial policy as being irksome. Smuggling (and privateering during the war with France), which was connived at by surprising segments of the colonial community, served somewhat to balance official restriction with unofficial enterprise.

Attempts by the Exchequer to impose direct taxation upon the colonies began in 1765 with the Stamp Act. The colonies protested and a Stamp Act Congress met in New York to issue a "Declaration of Rights and Grievances". Stamp agents were intimidated and forced to resign. Although Parliament repealed the Stamp Act in 1776, a crucial element of trust had broken down and a pattern of organized protest, intimidation and violence established. The idea that Parliament had the right to choke or even destroy colonial commerce raised the stakes.

> We have an old Mother that peevish is grown,
> She snubs us like Children that scarce walk alone:
> She forgets we're grown up and have Sense of our own;
> *Which nobody can deny, deny,*
> *Which nobody can deny.*
> Ben Franklin, "The Mother Country", *c.*1765.

Sir Alexander Mackenzie (1764–1820) was the first European to complete the northern overland journey across North America. He arrived in Canada from Scotland in 1779 and entered the employment of a leading Montréal fur merchant. Mackenzie served in Detroit, built trading posts in the far West, and led the expedition which followed the unknown Mackenzie River to the Arctic in 1789, and in 1793 reached the Pacific. His Voyages from Montréal ... through the Continent of North America to the Frozen and Pacific Oceans, *1801, contains a history of the Canadian fur trade.*

It was no longer just a threat to well-being which was feared. The Townshend Acts in 1767 imposed import duties upon glass, lead, paper and tea. A town meeting in Boston drew up a list of British products which Massachusetts vowed not to purchase. Other colonies followed the non-importation agreements. Merchants who declined to fall into line were roughly handled. The Townshend duties were amended in 1770, and the non-importation movement was abandoned. But feelings in the cities had grown to fever pitch, with

After landing 10,000 troops along beaches to the south of Veracruz in February 1847, General Winfield Scott (1786–1866), known to his contemporaries as "Old Fuss and Feathers", took Veracruz by siege, and marched 55 miles along the National Road until he encountered a Mexican force of 13,000 under Santa Anna in a fortified position at Cerro Gordo. In a stirring assault on 18 April involving hand-to-hand fighting, Santa Anna was routed, and Scott resumed the American march which ended with the capture of Mexico City in September.

regular conflicts between the British troops and the Sons of Liberty who provocatively erected liberty poles to taunt the redcoat garrisons. The battle of Golden Hill in New York on 19 January 1770, and the Boston massacre which followed on 5 March, became patriot *cause célèbres.*

Pressures from "below", and disorder in the streets, alarmed respectable men, loyalist or patriot. The leaders of the Boston Tea Party (16 December 1773) were not brawlers from the taverns, but disciplined men prepared and willing to oppose British taxation. Conservative patriots in New York saw the need to place "the Command of the Troops in the hands of Men of property and Rank who, by that means, will preserve the same Authority over the Minds of the people which they enjoyed in the time of Tranquility" (James Duane to Robert R. Livingston, Jr., 7 June 1775).

After further tea protests, Parliament responded with the Boston Port bill (31 March 1774), which closed Boston Harbor until the East India Company and the customs had been compensated for losses. Even more drastic legislation followed against Massachusetts, effectively revoking the colony's charter. Denouncing these measures as "coercive" and "intolerable", the colonies' protest took an unexpected turn. Instead of renewing the non-importation pacts, a Continental Congress met in Philadelphia (5 September 1774). Congress declared the Coercive Acts unconstitutional; Massachusetts was urged to form a government and collect taxes, but to withhold them from the royal government; the people were advised to arm themselves and form militias; and sanctions were declared against trade with Britain.

In Massachusetts, General Thomas Gage ordered the seizure of cannon and powder belonging to the province, and fortified Boston Neck (which would have isolated the city in the event of disorder or insurrection). In response, the Massachusetts House reconstituted itself as a Provincial Congress and appointed John Hancock to head a Committee of Public Safety empowered to call out the militia.

The prosperity of the Virginia planters in the colonial era, and their reputation for civility and culture, was often contrasted with the more commercial spirit to be found in the northern colonies. The aristocratic Virginian is represented in a trade sign, London.

On 19 April Gage ordered 700 men of the Light Infantry and Grenadiers under Lt. Col. Francis Smith to destroy the supplies which the province had stored in Concord. The movement of the troops crossing from the Common to Cambridge by boat was observed, and Paul Revere reached Lexington in time to warn Sam Adams and John Hancock. The British troops found a group of 70 militiamen opposing them at Lexington. They were easily dispersed, but by the time Smith reached Concord, the area was swarming with militia. The British column was harried along the entire route back to Charlestown, by which time they had taken heavy casualties: 73 dead, 174 wounded and 26 missing. The patriots suffered 93 dead, wounded or missing. The war that followed lasted until the defeat of Cornwallis at Yorktown in October 1781. A formal cessation of hostilities was not declared until January 1783.

Peace, again, and consequences

When peace was signed in 1783 at Paris, a new nation contemplated the sobering prospect of writing democratic constitutions, constructing a system of law, and through the Articles of Confederation, establishing the basic principles which were to govern relations between what had been largely autonomous and proudly individualistic colonies. Although American attempts at invading Canada failed, trade with Britain was disturbed by privateers. And the provinces most vulnerable to invasion experienced something of a civil war between "loyal" and "rebel" populations, which only ended when the "rebel" inhabitants departed south. Sympathy for the American cause was largely extinguished, and suspicion and fear of the Americans became a significant element in the emerging Canadian identity. After the war "loyalists" from the United States settled in Nova Scotia and Québec. (Benedict Arnold lived in St. John from 1786 to 1791). Once settled in Canada, the loyalists strengthened demands for a stronger role in local government. The most conservative elements in colonial American society found Canada somewhat too oligarchic for their taste.

Once again, the painters found work. The revolution helped Americans (or at least New Englanders) overcome their puritanical distrust of luxury and fine art. Patriotic sentiment (and new public buildings with large walls to fill) created a demand for heroic images of the making of the nation. Portraits of the revolution's leaders and heroes (a ceaseless hunger for portraits of the unsmiling Washington), representations of the signing of the Declaration of Independence, and a host of military paintings recorded the national apotheosis. America now began to have its own story.

The Crystal Palace in New York (1853), built upon the current site of Bryant Park on Sixth Avenue between 40th and 42nd Streets, was based on the design of the grander Crystal Palace built in London in 1851. It symbolized the handsome advance of trade and manufactures which enabled the former colony to challenge the supremacy of Britain in the world's markets. By design the Crystal Palace paid respectful homage to London, but in spirit it was an assertion of New York's self-confidence and economic power.

American War of Independence

The long struggle for independence against British rule enabled the thirteen colonies to forge a national identity.

"We shall be in the fort in two minutes." The last words of Brigadier General Richard Montgomery to his aide-de-camp Captain Aaron Burr, during the failed assault on Québec, 31 December 1776.

Benjamin Franklin, London agent of the Massachusetts legislature, overheard a British officer boasting in the spring of 1775 that "with a thousand British grenadiers, he would go from one end of America to the other, and geld all the males, partly by force and partly by a little coaxing." British arrogance and ignorance of the character of the colonials, allied to ever-increasing taxation and the "intolerable" laws passed to enforce British policy, deepened mutual distrust.

When General Thomas Gage attempted in 1775 to confiscate supplies held by the Massachusetts Provincial Congress in Concord, he found militia and irregulars willing to oppose the redcoats with arms. At Lexington on 18 April, a skirmish left eight Americans dead. After coming under attack at North Bridge at Concord, the British withdrew. Massachusetts militiamen poured fire into the column along the 21-mile route back to Boston. The British lost 273 men killed. There was no more boasting about gelding American males.

In the war that followed, the British had the advantage of the Royal Navy, but their leadership was seldom able to exploit the weaknesses of the patriots. When they confronted the Americans at battles like Germantown, Pennsylvania, and Charleston, South Carolina, superior British discipline prevailed, but the struggle to win over the uncommitted was weakened with every colonial casualty. The American commander, George Washington, was realistic, unflappable, and uncompromisingly devoted to the cause of independence. The patriots lost battles, but they endured. Word that the French fleet under Admiral de Grasse would arrive at Chesapeake Bay on 13 August, 1782, persuaded Washington to launch a joint attack with French forces led by General Rochambeau. They besieged General Cornwallis at Yorktown, blocked the attempted British reinforcement by Admiral Graves, and won the decisive battle of the war. Yorktown assured American victory and independence.

The surrender of 8000 men under Cornwallis at Yorktown on 19 October 1781 was as decisive a battle as the victory of Wolfe at Québec in 1759. It ended all British thoughts of victory, and with the fall of the North ministry in Westminster in March 1782, the way was open for peace and American independence.

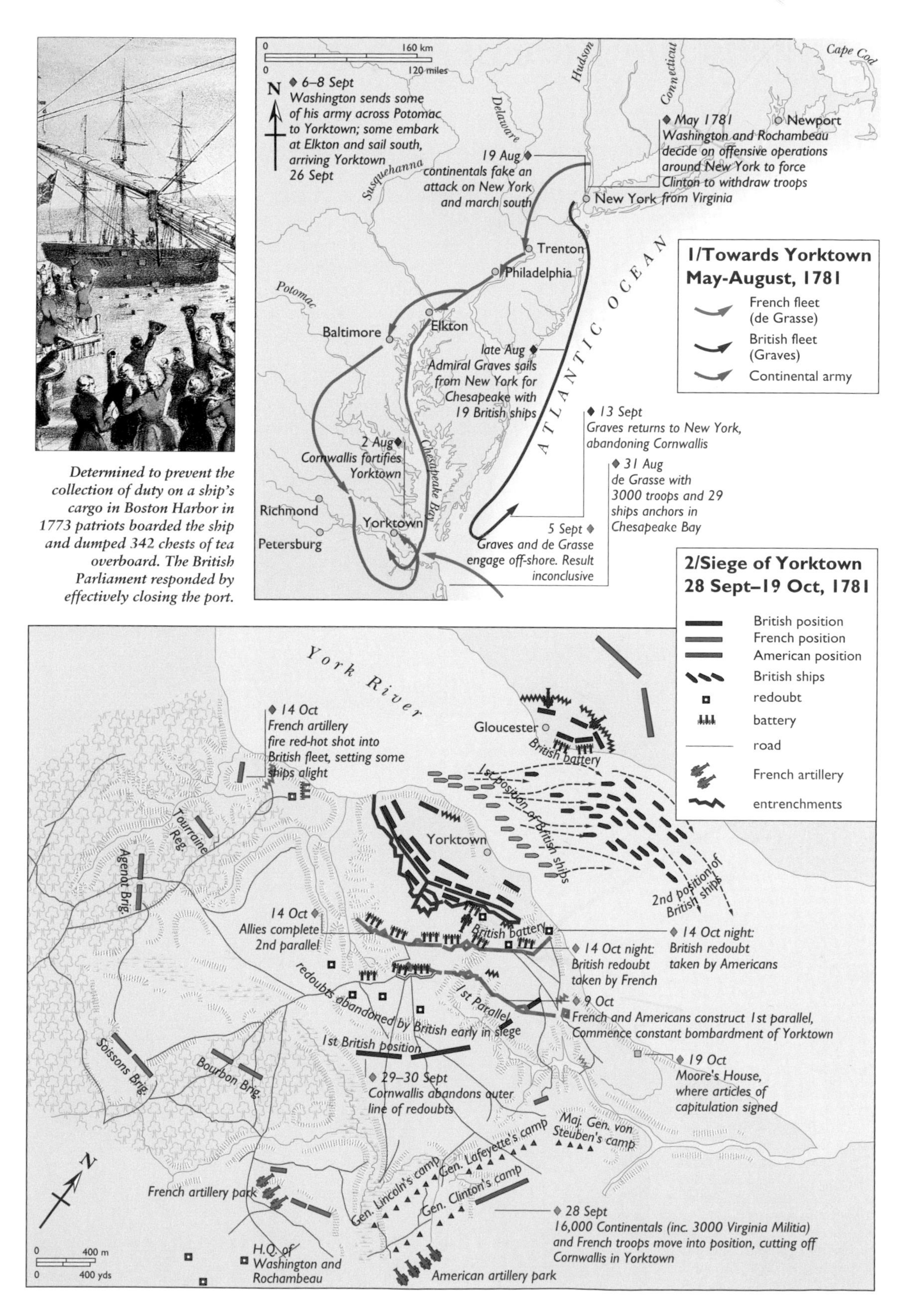

Determined to prevent the collection of duty on a ship's cargo in Boston Harbor in 1773 patriots boarded the ship and dumped 342 chests of tea overboard. The British Parliament responded by effectively closing the port.

Towards the Louisiana Purchase

Territorial disputes and the writing of a new constitution preoccupied the new nation for a generation.

"We were confined within some limits. Now, by adding an unmeasured world beyond that river [the Mississippi], we rush like a comet into infinite space."
Fisher Ames, Federalist of Massachusetts, on the Louisiana Purchase, 3 October, 1803

Peace negotiations with Britain after Yorktown proved drawn-out and complicated, continuing well into 1784. In the meantime, the revolutionary army demobilized and went home. The new states, loosely bound together by the Articles of Confederation, retained considerable autonomy, and could not be forced to accept the terms negotiated with Britain. As a result, redcoats remained in occupation of Detroit and other western forts until pre-Revolutionary British debts and other claims were settled. Undisputed territory north of the Ohio River was organized by the Northwest Ordinance, passed by the Congress of the Confederation on 13 July 1787. Slavery was to be excluded, and when the population had grown, new states would be admitted to the union on an equal basis.

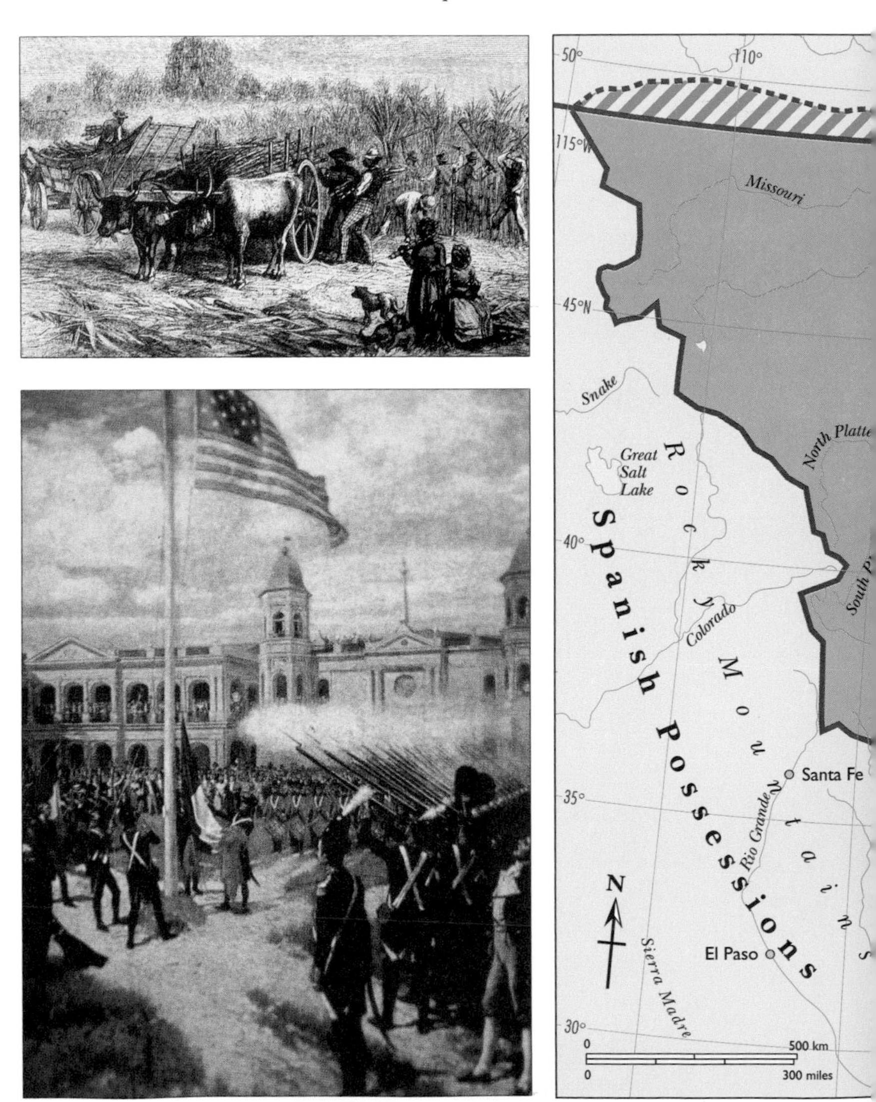

Slavery and sugar were intimately connected across the world. The Louisiana sugar plantations were among the most productive and profitable in the United States. Slaves are seen cutting and loading sugar cane ashore.

A salute is fired to celebrate the raising of the Stars and Stripes in New Orleans, 1803. The purchase of the Louisiana Territory from France doubled the area of the United States and finally removed the spectre that a European power would close the Mississippi River to Western commerce.

It was a happy time for lawyers, with constitutions to be written for all the states, and legislatures to be established. International relations proved more troublesome; Spain refused United States' ships free navigation on the lower Mississippi, and there was no agreed border with Spanish Florida.

The inadequacy of the Articles of Confederation led to a convention held in Philadelphia to rewrite the Constitution. The resulting Constitution, ratified in 1788, established representation in the lower house by white population plus three-fifths of the black population.

John Jay's treaty in 1794 secured British withdrawal from the Northwest posts and settled the outstanding trade disputes, but the western border of the United States was still uncertain after 1800, when Spain ceded the Louisiana Territory back to France. Napoleon's decision to abandon for a time French colonial adventures in the New World opened the way for the sale of the territory to the United States. For 60,000,000 louis ($15,000,000) President Jefferson expanded the size of the United States by 828,000 square miles. Formal possession was assumed on 20 December, 1803. Westward expansion now had no political constraint.

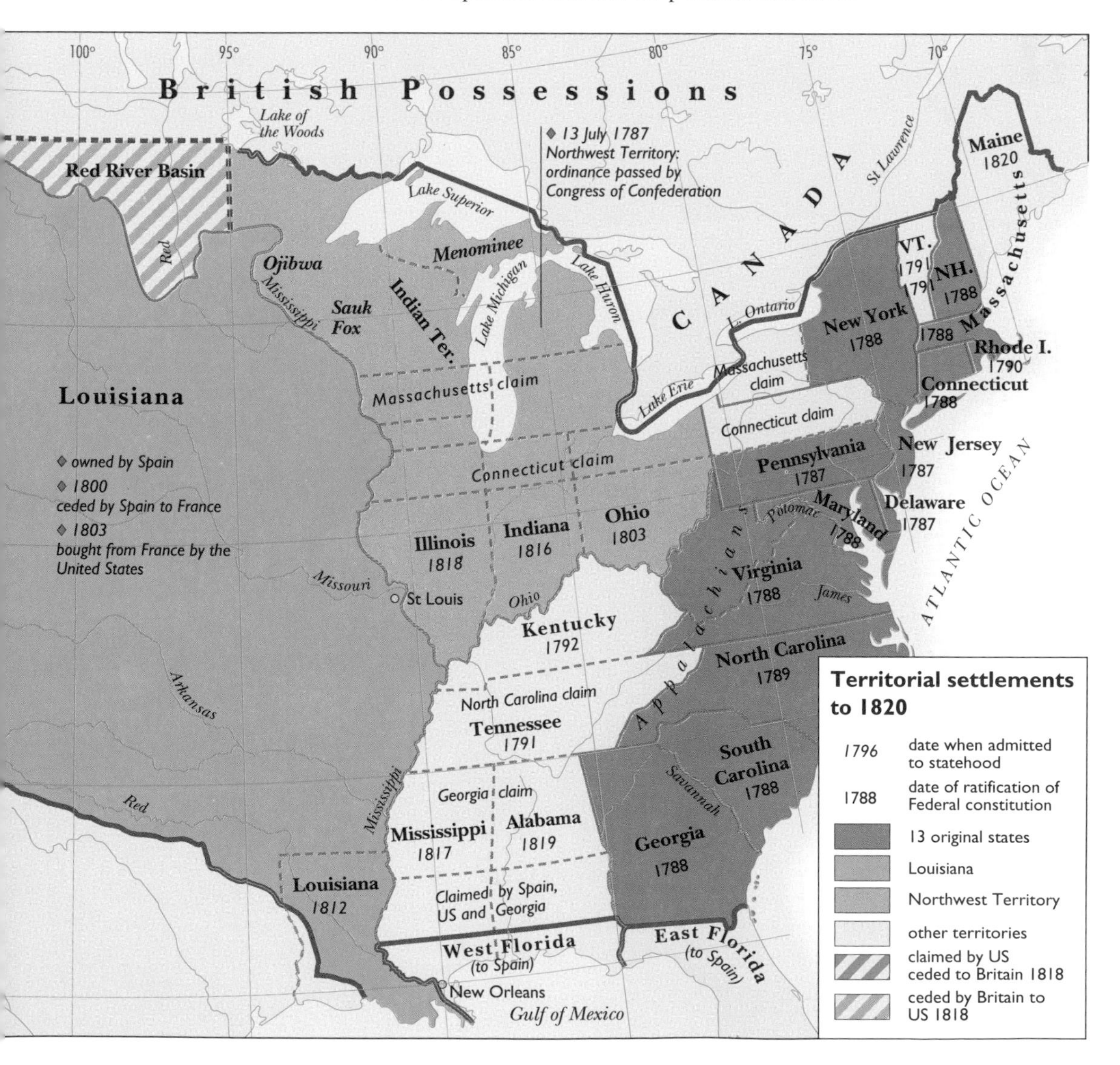

Canada to the War of 1812

Thirty years after Yorktown, the European war against Napoleon once again drew the United States and Britain into armed conflict.

> *"The Conquest of Canada is in your power. I...believe that the militia of Kentucky are alone competent to place Montréal and Upper Canada at your feet."*
> Henry Clay, Speaker of the House, in Congress, 1812.

The Treaty of Paris in 1763 confirmed British rule from Hudson Bay to the border of the Spanish empire. For thirteen years the British were preoccupied with the stubborn resistance of the 13 colonies, and the governance of the Canadian territories newly acquired from France. The majority of the population were *habitants*, French-speaking Roman Catholics, a religion subject to a severe lack of civil liberties in Britain. British seizure of the Dutch colony of New Netherlands in 1664 suggested the wisdom of a policy of tolerance and co-optation of the local élites in the system of British rule.

New Englanders had long coveted the former French fishing grounds off Canada, and when war broke out in 1775 the newly-formed Continental Congress sought to strike at British forces in Canada. The defeat of the American expedition at Québec in 1775, and the total failure of a British invasion of New York state mounted by "Gentleman Johnny" Burgoyne in 1777, offered lessons—soon forgotten—on all sides.

Both states (Britain and USA) settled back to peace, westward expansion and the development of internal government. The arrival in Canada of thousands of American-born loyalists strengthened pressures for the creation of local assemblies and democratic institutions.

The British naval blockade of Napoleonic France infringed upon the rights of neutral American shipping, but President Jefferson's trade embargo of 1807–09 failed, and intense American indignation led to a declaration of war on Britain in 1812. For two years ill-trained American militia, incompetent officers and under-equipped regulars attacked Canada, and achieved little. A pause in the war with Napoleon in 1814 enabled the British to plan

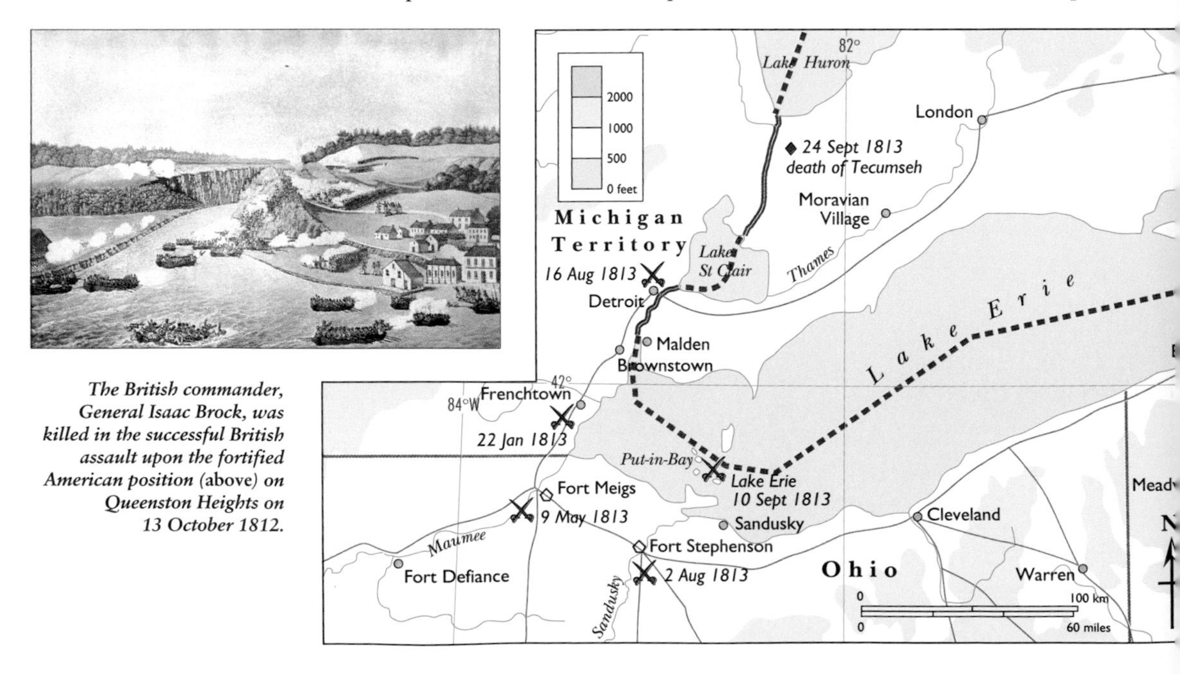

The British commander, General Isaac Brock, was killed in the successful British assault upon the fortified American position (above) on Queenston Heights on 13 October 1812.

a sweeping attack from Lake Champlain and from New Orleans. Washington, D.C. was captured and the White House burnt in August. But American naval victories on Lake Champlain turned back the invasion. Andrew Jackson's bloody victory at New Orleans blotted out memories of military failure. Peace was signed at Ghent, Belgium, leaving commissions to settle the remaining disputes. An unnecessary war, fought with bravery, cooled dreams of military conquest in the north for good.

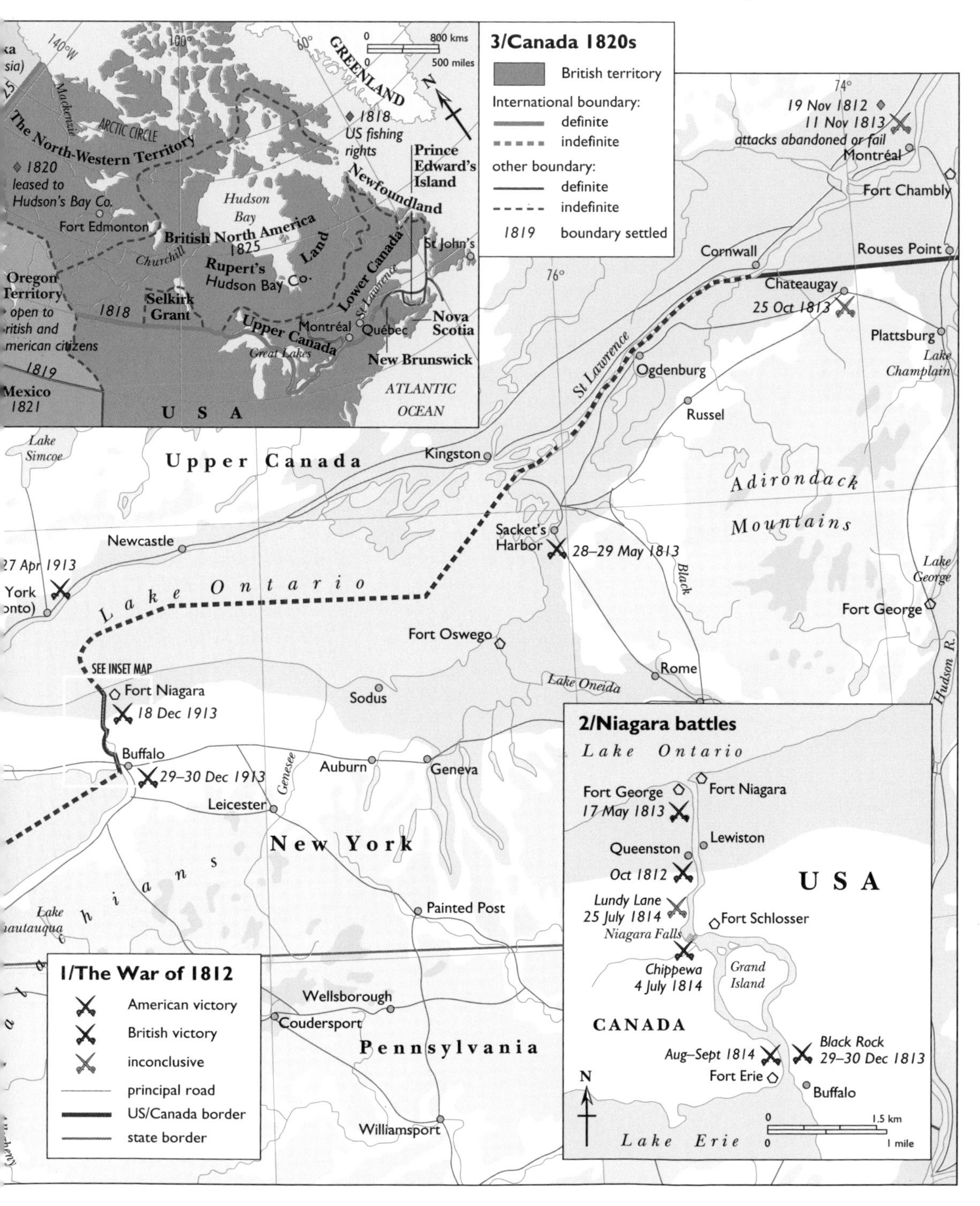

Mexican Struggle for Independence

Defending their interests, conservatives in Mexico declared independence from Spain in 1821.

"The plaza and the streets were littered with broken pieces of furniture and other things robbed from the stores, of liquor spilled after the masses had drunk themselves into a stupor."
Lucas Alamán, an eyewitness of the sack of Guanajuato, 28 September 1810

There were tensions in New Spain between the *criollos* (of Spanish descent, long established in Mexico, who had begun to see themselves as a distinct people), and the *peninsulares* (Spaniards occupying leading positions in the colonial administration and the church), and resentment of taxation and state monopolies. Expulsion of the Jesuits in 1767 deepened the disaffection of the *mestizo* (of Indian-Spanish descent) priests and Indians. Old grievances were long remembered.

A minority of *criollos* were moved by the American War of Independence and Declaration of Independence in 1776, and the French Revolution of 1789, and hesitantly asked questions about the rigid Spanish imperial rule. Fear of the Indian masses restricted to a minority the appeal of Enlightenment ideas about the rights of man. Most *criollos* wanted freedom but feared the consequences of social change.

Napoleon's intervention in Spain in 1808 destroyed dreams of peaceful change. A parish priest, Miguel Hidalgo y Costillo, had discussed Mexican independence with a small group of *criollos* in Guanajuato. Forced to launch a revolution prematurely, Hidalgo found the poor ready to rise against Spanish rule. But sacked towns and murdered Spanish garrisons drove wealthy *criollos* to abandon the cause of independence. The insurrectionaries were defeated and executed in 1811.

A *mestizo* parish priest and former student of Hidalgo's, José María Morelos y Pavón, continued the struggle for independence, seizing large territories from 1812 to 1814. But Morelos, too, was defeated by royalist troops and executed in 1815.

The promulgation of a liberal constitution in Spain in 1820 alarmed Mexican *criollo* conservatives, who plotted a break with Spain to preserve their position and that of the Church. Augustin de Iturbide, a *criollo* military officer, persuaded the rebels that the army shared their desire for Mexican independence. The *Plan de Iguala* of 1821 guaranteed the independence of Mexico as a constitutional monarchy; Roman Catholicism as the state religion; the abolition of slavery and equal treatment of *criollos* and *peninsulares.*

Weary of a decade of struggle, the people rallied to a conservative plan for independence. Iturbide reached Mexico City on 27 September, 1821. A governing junta was established the following day.

The Metropolitan Cathedral and the Sagrario (Sacrarium, right***), stand on the Zócalo in the historic centre of Mexico City. A cathedral on this site was begun in 1525. The Sagrario was built in the 18th century.***

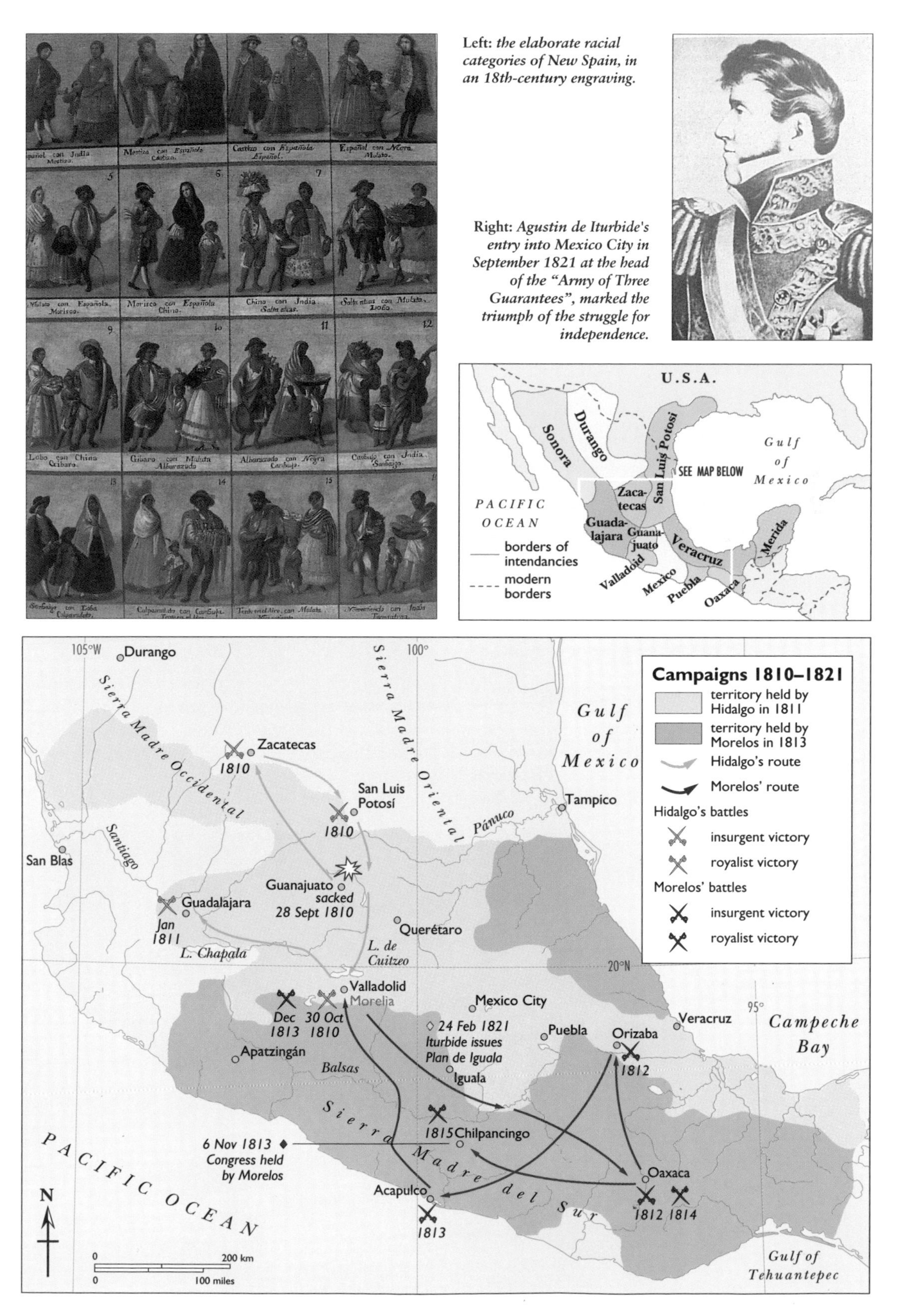

Left: *the elaborate racial categories of New Spain, in an 18th-century engraving.*

Right: *Agustin de Iturbide's entry into Mexico City in September 1821 at the head of the "Army of Three Guarantees", marked the triumph of the struggle for independence.*

Texas and Expansion

The struggle for liberation in Texas in 1836 led to war between Mexico and the United States in 1846.

"The Annexation of Texas ... is the first step to the conquest of all Mexico, of West India ... of a maritime, colonizing, slave-tainted monarchy, and of extinguished freedom."
From the diary of John Quincy Adams, 10 June 1844

The land hunger of the population of the southern territory of the United States concerned the authorities in New Spain. They hoped to strengthen Spanish control by admitting colonists on a restricted basis. A charter was granted to Moses Austin in 1821 to bring 300 Catholic families into the territory. Land was sold for 10¢ an acre. By 1835, there were 30,000 *gringos* living in Texas (a northern province of Mexico), both whites and slaves.

After Mexican independence in 1822, the presidency changed hands 36 times between 1833 and 1855. Mexico was unstable and divided, and further colonization was banned in 1830. The loss of political representation —when the liberal constitution of 1824 was replaced by a more autocratic document in 1836—angered the gringos, who demanded independence.

In the belief that a few salutary massacres would teach the Texans a lesson, the Mexican Emperor Santa Anna besieged the mission at the Alamo, and had all of the surviving defenders killed. More than 300 Texas troopers were executed on Santa Anna's orders at Goliad. But on 21 April 1836, Santa Anna's column of 1200 men was routed by General Sam Houston at the San Jacinto River, thus securing Texan independence. Annexation of Texas by the US was delayed by fears that another slave state would destroy the political balance of the Union. Annexation was not achieved until December 1845.

American expansionists looked for a pretext to attack the disunited Mexicans. A border clash led to a declaration of war on 12 May 1846. Despite heroic resistance, an easy United States victory was sealed by the Treaty of Guadeloupe Hidalgo. The border was located on the Rio Grande. Mexico lost two-fifths of her territory, and received $15 million as an indemnity from the jubilant Americans, who had secured California for a pittance.

Samuel Chamberlain's "Battle of Buena Vista" portrays the victorious commander of the US troops, General Zachary Scott, on a white horse. The Mexicans under Santa Anna attacked an entrenched US position on the Saltillo–San Luis Potosi road on 22–23 February 1847. Scott's victory ended the war in northern Mexico.

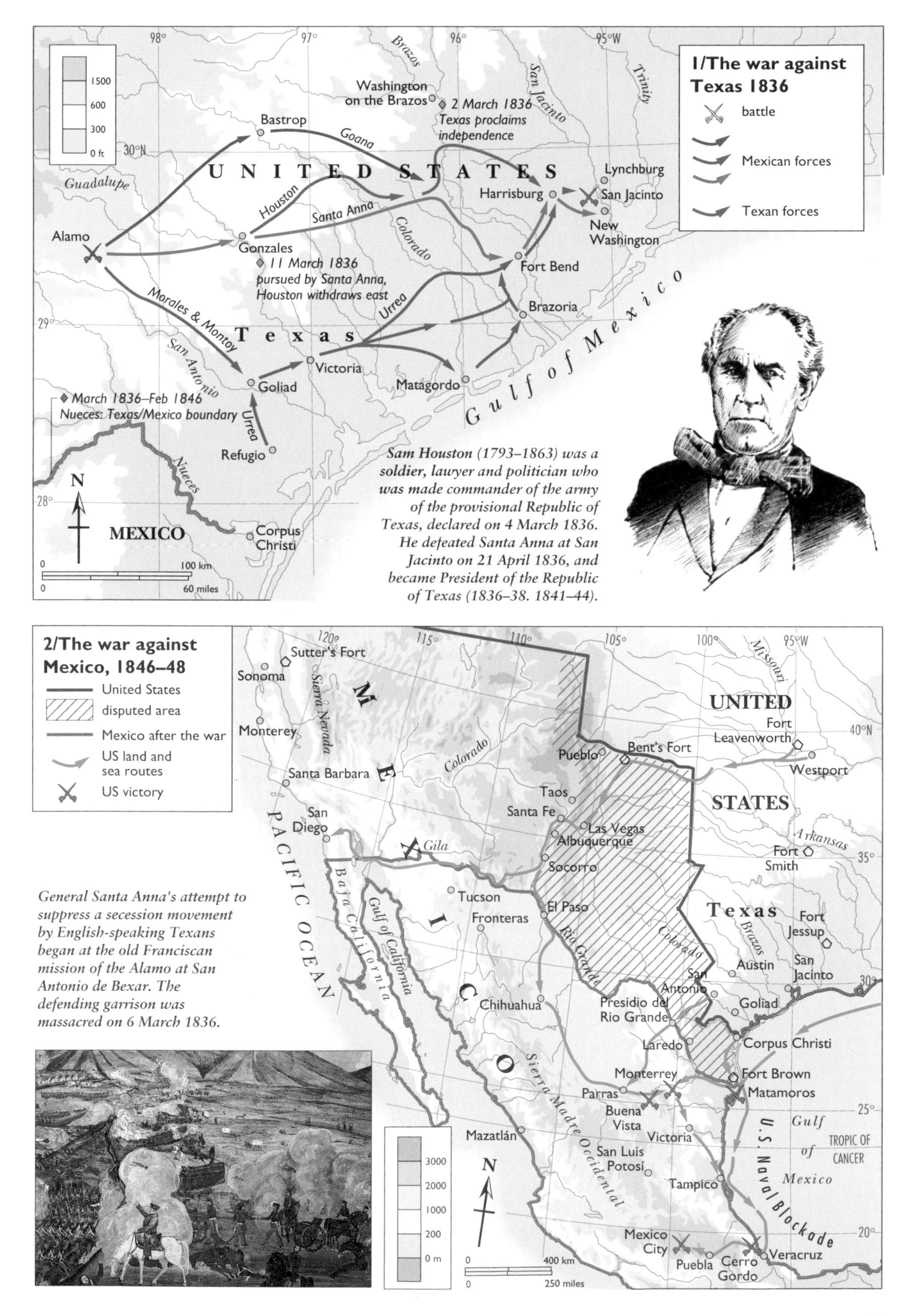

Sam Houston (1793–1863) was a soldier, lawyer and politician who was made commander of the army of the provisional Republic of Texas, declared on 4 March 1836. He defeated Santa Anna at San Jacinto on 21 April 1836, and became President of the Republic of Texas (1836–38. 1841–44).

General Santa Anna's attempt to suppress a secession movement by English-speaking Texans began at the old Franciscan mission of the Alamo at San Antonio de Bexar. The defending garrison was massacred on 6 March 1836.

Canals, Railroads and Cotton

New methods of transportation transformed the economy of the United States.

"A profitable investment ... was not the moving cause for the investment ... [Railways] were constructed for the promotion of the interests of their respective States and Cities where they terminate."
John Wood Brooks to Erastus Corning, 19 December, 1859

With the construction of the Erie Canal (350 miles from Albany to Buffalo, opened in 1825), the transportation "revolution" began which transformed the American economy. The price of bulk goods carried by land fell 95 per cent between 1825 and 1855. Rural farmers entered the national and international market. Regional self-sufficiency declined.

Canals were well-suited to the transport of coal from the anthracite regions of Pennsylvania, and railway lines and steam-powered locomotives equally served specific, local purposes. Ellicott's Mill was connected to Baltimore by a 13 mile railway line in 1830. Railway lines were a form of regional economic warfare, drawing trade away from commercial rivals. Boston and New York, Savannah and Charleston, Baltimore and Philadelphia fought fierce battles for commercial supremacy. By the 1850s, the railway had largely supplanted the canal in the United States economy.

Entrepreneurs believed that the railway was an "iron civilizer", converting sleepy villages into thriving commercial centres. Growth and prosperity clearly lay in the new tranportation networks. The railway was also an expression of modernity, of national pride. Begun as early as 1837, the line from Veracruz to Mexico City, which rose from sea-level to over 9,000 feet, was the most daunting engineering challenge on the continent. But Canada, with only 60 miles of railway in 1850, and Mexico with barely 150 miles of track a decade later, were dwarfed by the 30,000 miles of track in the United States.

Southern planters invested in railway stock and slaves worked to build the lines, but the voracious appetite for good cotton land, and the speed with which it was exhausted, made investment in extensive rail links unattractive. Two-thirds of the United States rail network lay in the northern states, giving the Union Army an industrial and military advantage when the Civil War broke out in 1861 which the Confederacy could not match.

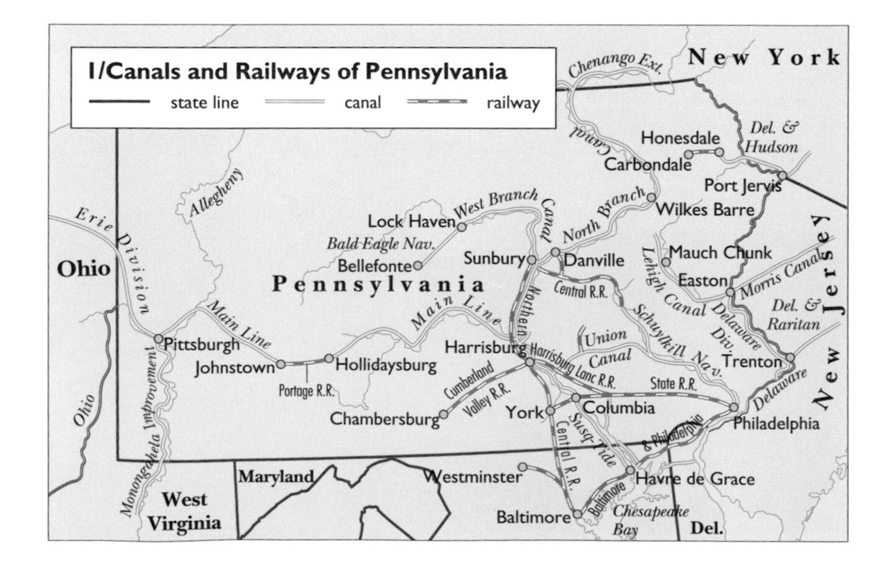

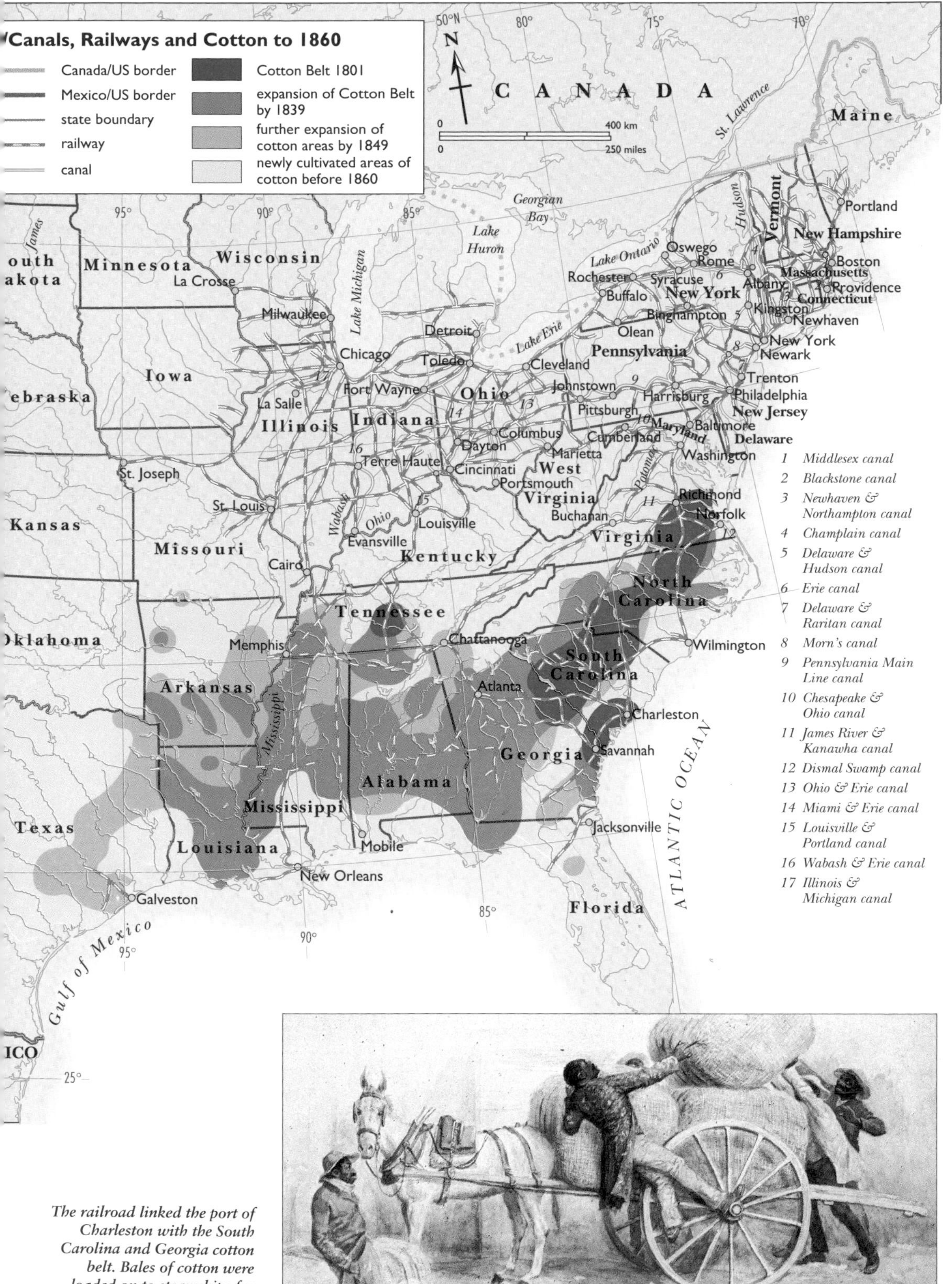

The railroad linked the port of Charleston with the South Carolina and Georgia cotton belt. Bales of cotton were loaded on to steamships for shipment to New York and Liverpool.

Gold and California

Immigrants from the United States seized California in 1846, and within two years the world's richest goldfield was found in the Sacramento Valley.

The Spanish explored the coast of *Alta California*, and sent military expeditions to take possession of the territory to the north of their Mexican domain. The *Presidio* (military garrison) was established at San Francisco in 1776. With the soldiers came the Franciscan friars, led by Fr. Junípero Serra, who had ambitious plans to convert the Indians. Twenty-one Franciscan missions were established in California.

Other powers were beginning to show an interest in California: a trading vessel sent by Boston merchants arrived in Monterey in 1796. The Russian-American Company, hunting sea-otters, built a stockade at Fort Ross. (Like the Dutch settlement at Albany in the 1620s, all inhabitants were company employees.) The Spanish claimed exclusive title to California, and resisted the Russian presence.

As relations worsened between the United States and Mexico, American settlers staged a coup at Sonoma on 14 June, 1846. By the time General Stephen Kearny's column arrived in California from Fort Leavenworth, he found Americans firmly in control.

On 24 January 1848 a worker near Sutter's sawmill at Coloma on the American River discovered a lump of gold. As word of the find raced across the territory, and then around the world, California became the object of a torrent of migrants, speculators, drummers, frauds, desperadoes and land-hungry settlers. Half a billion dollars of gold was taken from the California goldfields in five years.

The California constitution of 1849 banned slavery, and its admission into the United States as a "free" state was strenuously resisted by the slaveholders. As part of the great "Compromise" of 1850, the South obtained the Fugitive Slave Law, and California was admitted to the Union.

Top right: *the Santa Clara Mission was at the centre of extensive vineyards and grazing lands which made the system of missions a powerful force in the growth of Spanish civilization in California.*

Right: *The discovery of gold at Sutter's Mill on 24 January 1848, gripped the imagination of people across the globe. This French engraving gives a panorama of the process from extraction to the sale of the gold dust.*

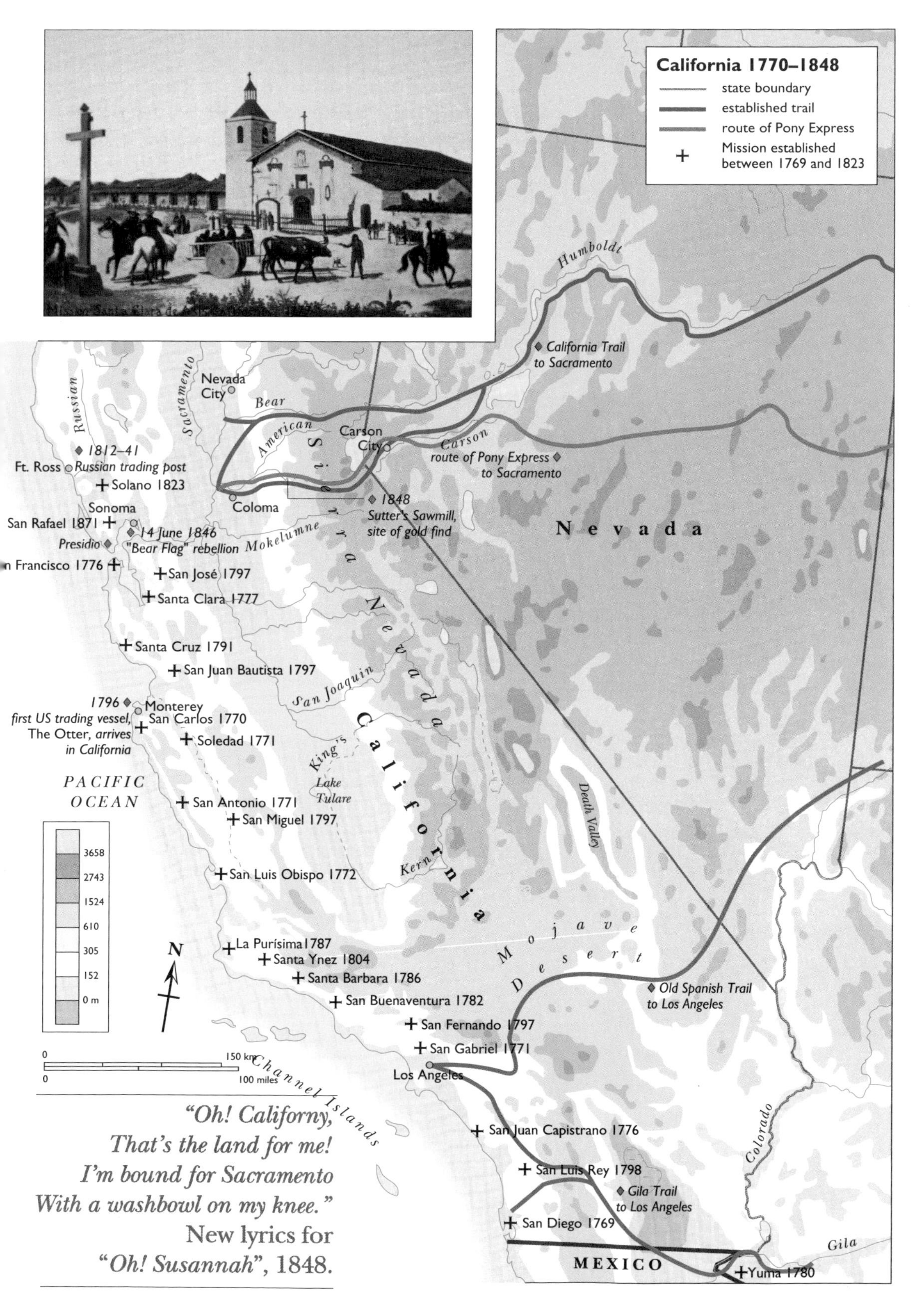

"Oh! Californy,
That's the land for me!
I'm bound for Sacramento
With a washbowl on my knee."
New lyrics for
"Oh! Susannah", 1848.

'An Outrage on American Womanhood'

New definitions of the American woman struggled to keep pace with the changing experience of women in American life.

"I ain't a young lady."
Jo March in *Little Women*, 1868

"The American girl" was invented by the novelist Henry James during the winter of 1878. He had been living in Italy the previous year, and wrote a little tale about a young American girl, Daisy Miller, cast adrift in the ambiguous complexities of expatriate life in Italy and Switzerland. The first American editor to read the story rejected it without comment. James was later told that *Daisy Miller* was seen as "an outrage on American womanhood". An English periodical, less sensitive about American feelings, published it. James, an "amiable bachelor" by his own description, had told the story of a young American girl destroyed by her innocence, and by the failings of a society in which there were no rules, standards or ancient traditions. Ill-mannered American boys, who spoke freshly to adults, and adolescent girls already familiar with "gentlemen's society" were ingredients in the making of what James regarded as a "little tragedy". American innocence was a dangerous thing, he felt.

Were these ill-mannered and indisciplined creatures the American Woman? Many articles and books hotly argued that they were a caricature of the truth. But the great imagined woman in American culture, like Hester Prynne (in Nathaniel Hawthorne's *The Scarlet Letter*, 1850), was a scorned adulteress. The popular heroine of schoolgirl fiction, Jo March (in Louisa May Alcott's *Little Women*, 1868), was a tomboy who rejects the love of Laurie. Transgression, discontent, rebellion and a dangerous innocence: such were the sins of femaledom in the American imagination in the 19th century.

There had been significant changes in the experience of American women since the colonial era. By far the most important emerged in the 1830s with the idea that there was a "woman's sphere" centred on the home. A "cult of domesticity" spread across America. It represented a gain in authority, in which women tended to displace their husbands in domestic management and responsibility for their children's education and moral upbringing, while sharpening the distance between the effective realm of women and men who worked away from the home and were distant, pressured deities.

Bottom left: *The sewing machine was the first piece of new technology which was widely used by American women. The Wheeler and Wilson machine, as advertised in 1858, could, as the accompanying illustration suggests, be introduced into the domestic sphere without threat to feminine modesty.*

Top left: *A Progressive-era couple, 1906: he has a copy of the muck raking* McClure's Magazine *under his left arm, and she has a handsome parasol on her shoulder. Taken in Cornwall, Connecticut, by T.S. Bronson.*

Top right: *A moment for a cigarette and mug of beer by the fire, Washington D.C., caught by Frances Benjamin Johnston, who became one of America's first photo-journalists.*

For this task, the housewife had the new "domestic science" to help plan and organize the home. But the laws which made a married woman little more than her husband's chattel remained in force. On marriage, women lost their name and the right to their property. They could not vote, nor enter into binding legal contracts. In the pages of *Godey's Ladies Book* or *The Ladies' Home Journal*, women were urged to become more refined in manner, more conscious of social position, more pious and regular in church attendance, more obedient to their parents and husband, more self-sacrificing before the needs of their offspring.

Women were hemmed in by stiff whalebone corsets and sententious advice. They were accompanied by chaperones, patronized by men and denied anything more than the education deemed appropriate to the female station. The judgement American matrons made of Hester Prynne, Jo March and Daisy Miller was harsh. In drawing rooms across the nation, there was a swelling anti-Daisy Millerite sentiment, and no little indignation at the false picture of American women which James had given the world. It was their daughters (and granddaughters) who secretly rejoiced in Hester's strength, Jo's burning ambition, Daisy's rebelliousness, seeing perhaps something of their future take shape, chapter by chapter.

Right: *A turn-of-the-century soda fountain selling notions (watch straps, hair dye, shoe polish, laxatives), and an array of flavoured drinks, taken in Springfield, Massachusetts. by George and Alvah Howes.*

IV: Spanning a Continent

It was an age of self-made men, but the steam-engine and the engineer defined the achievements—and the failures—of America after the Civil War.

"There's an honest graft, and I'm an example of how it works. I might sum up the whole thing by sayin': 'I seen my opportunities and I took 'em.'"
George Washington Plunkitt, 1905

In an address to the 148th Ohio Regiment, delivered in Washington on 31 August 1864, President Abraham Lincoln argued that:

> "Nowhere in the world is presented a government of so much liberty and equality. To the humblest and poorest amongst us are held out the highest privileges and positions. The present moment finds me at the White House, yet there is as good a chance for your children as there was for my father's."

Since the era of Andrew Jackson, the belief that people of humble background could rise to the highest positions in American society had become a matter of pride, virtually a pillar of the American republic. Republicans and Democrats, Native Americans and even the more enlightened Whigs agreed at least on this: America offered high opportunity to the "humblest and poorest".

It was also accepted, and sometimes explicitly, that these opportunities were generally inappropriate for women. Male Americans who had achieved

After the end of Reconstruction in the 1870s, Americans increasingly accepted a reinterpretation of the Civil War that was much more sympathetic to the South. As the North became industrialized, the Southern plantation came to stand for a more gracious way of life. Slavery was marginalized in a process which culminated in the 1939 film version of Margaret Mitchell's Gone With the Wind. *Rhett Butler, played by Clark Gable, discusses the price of cotton with Scarlet O'Hara, played by Vivien Leigh.*

great things showed considerable pride in having risen from humble beginnings. Lincoln, in autobiographical notes written during his campaign for president in June 1860, adopted the third person to describe his schooling: "A. now thinks that the aggregate of all his schooling did not amount to one year. He was never in a college or Academy as a student; and never inside of a college or academy building till since he had a law-license." An obituary of Cornelius Vanderbilt, published in the New York *Herald* in 1877, remarked of Vanderbilt's background that "he was always a wild young fellow, who knew more about hoeing weeds out of his mother's cabbage patches than he knew of the addition table. The times, and the customs of the class from which he sprung, did not make him a lover of books..." The inventor of the revolver, Samuel Colt, "was always so full of restless energy that he greatly preferred to be working in the factory to going to school."

Family circumstances, a highly mobile society, and the modest schooling available in most communities, combined to limit opportunities for further study. After the bankruptcy and death of his father, Herman Melville was withdrawn from Albany Academy in 1832 at the age of 13 to become a clerk in an Albany bank. Many Americans felt that this was not necessarily a misfortune, and American values often elevated practical skills above book learning. The young man, of however lowly a background, who had the right attitude and who worked hard, embodied a social ideal which won broad acceptance.

Cornelius Vanderbilt (1843–1899), grandson and namesake of Commodore Vanderbilt, and son of William H. Vanderbilt, was the very model of an elegant, upper-class gentleman whose sartorial model was English.

Young men were taught these values in thousands of sermons, in chats across the domestic hearth, and in a growing literature specifically aimed at younger readers. Among the authors of these tales, Horatio Alger, Jr., a former Congregational minister, achieved the widest audience with *Ragged Dick; or, Street Life in New York* (1868). So potent were Alger's tales that they were read by immigrant children on the lower East Side and children growing up in rural Midwest with as much enthusiasm. One such, who read *Ragged Dick* at the turn of the century, recalled that in Alger's books:

> "There were no divisions of 'classes' and 'masses'. There were rich and poor, but the rich could suddenly become poor, and the poor gradually became rich. It was a country of limitless opportunity for the moral, the virtuous, and the industrious. The poorest, the obscurest boy could aspire to the richest, most beautiful girl. As I read my very first Alger book, I fancied that the author was writing about me..."

After an unfortunate business involving young boys in his Congregational parish in Brewster, Massachusetts, Alger resigned the ministry and eventually reappeared in New York, attending services at Five Points mission located in the city's worst slum district. Alger based *Ragged Dick* on a street-wise youngster he met at the Newsboys' Lodging House in Five Points. Central to the success of the book was the character of Ragged Dick, an orphan living on the city streets. Dick is far from being a model boy, but his "frank, straight-forward manner", and determination not to do anything mean or dishonourable, wins our trust. Despite having few social or economic advantages, Dick was a straightforward, self-reliant, generous young person whose desire to "get on" in life sets him apart from the other young boys who are living on their wits. He possessed an aptitude, a willingness, for hard work. Alger equates getting on with learning to be "spectable". The other great adolescent in American literature, the hero of Mark Twain's *Huckleberry Finn* (1884), draws the opposite conclusion about "sivilization", despite the best efforts of Judge Thatcher and Miss Watson. Alger rather than Twain was the preferred author in many a respectable household.

Mark Twain, photographed in a rocking chair at his home in Hartford, Connecticut, embodied the American ambivalence about technology. He invested (and lost) a fortune on an innovative linotype machine, but sceptically feared that human nature was incapable of escaping from its atavistic prejudices and instincts to truly benefit from the new industrial possibilities of the age.

In the world of Ragged Dick, energy and industry are rewarded and indolence suffers. The development of desirable personal habits ("avoid extravagance and save up a little money if you can") is seen as the key link between (financial) rewards and character. The rewards are comically rapid: a dishonest clerk is dismissed on the spot, and his job offered to Dick. Above all, there is the message of Dick's friend and patron, Mr Whitney, about self-help: "Remember that your future position depends mainly upon your self, and that it will be high or low as you choose to make it... You know in this free country poverty in early life is no bar to a man's advancement." Alger's *Ragged Dick* embodies a crucial version of the values securely at the centre of American life after the Civil War.

In 1876, when Americans came to mark the centennial of their independence, it was both the rewards of energy and of industry which were celebrated. Congress authorized an "International Exhibition of Arts, Manufactures and Products of the Soil" to be held in Philadelphia (where the Declaration of Independence was signed). Despite the Depression which arrived in the fall of 1873, and gripped the economy until 1879, 51 foreign governments accepted the American invitation. A 284-acre site in

Fairmont Park was planted with 20,000 trees and shrubs. Miles of walks and surface railway lines were laid. A total of 249 buildings were erected. On 10 May President and Mrs Grant were joined on the grandstand by the Emperor and Empress of Brazil, the first reigning crowned heads to visit the United States. (It was commented upon that the President wore a business suit and no gloves.) A *Centennial Inauguration March*, commissioned from Richard Wagner, was played. A hymn, with words by John Greenleaf Whittier, was sung. After a speech by the President, and a hoisting of national banners, the President and the Emperor turned two silver-plated cranks which set in motion the machines which filled the Main Building. The crowd of 186,672 (including 110,500 admitted free) cheered.

Despite an admission charge of 50¢, attendance figures rose steadily. The Centennial Exhibition was instructive and patriotic. Some 13,720 exhibitors and perhaps a million objects on display gave the public value for money. Above all, Machinery Hall captured the interest of the press and visitors. At

Daguerreotype photography made a stunning impact in the United States in the early 1840s. It was a technology which required little formal education, and only a minimum of training. Daguerreotypists were drawn from proprietors of drug stores, sellers of notions, restauranteurs, and travelling salesman. Advances in technique (wet-plate collodion photography from the 1850s) made pictures easier to take. Equipment became lighter and more portable, exposure times were sharply reduced. The professional photographer (seen right *adjusting the aperture setting of a glass-plate bellows camera on a tripod) struggled to keep ahead of the democratic enthusiasm which made the "Kodak" and photography one of the first of the widespread national hobbies which required equipment which could not be manufactured at home.*

its centre was George Corliss's 56-ton steam-driven flywheel, which silently turned the shafts which set hundreds of lathes, drills, pumps and saws into motion. Water was pumped in arcing jets, wheels turned, fans revolved. Twice a day Machinery Hall was alive with the new energy of the age. The Corliss Centennial Engine became the most widely reproduced symbol of the Centennial. It was a symbol of American prowess and growing industrial might. (Its subsequent fate, to be sold in 1910 for scrap at $8 per ton, was no less symbolic: in the age of electricity even the most spectacular steam engines were unwanted junk.)

Technology and invention were to be found everywhere in Fairmont Park. There were steam-powered forging hammers, water-tube boilers, internal combustion engines, steam elevators, Pratt & Whitney metalworking equipment, portable steam engines to supply power for threshers, and cylinder presses. The Western Union company displayed the most advanced relays, keys and insulators for the telegraph. A whole building was set aside for photography.

In the Women's Pavilion a young lady was at the controls of a steam engine which powered a row of looms and a printing press. Two dozen women were present who had been awarded patents for a variety of inventions: from a gauge used to cut lozenges to a model house made of interlocking bricks. Patented garments, dressmaking systems and improved kitchen tools were on display.

The Centennial Exhibition in Philadelphia announced something a great deal more than the anniversary of the independence of the United States. It announced that the steam-engine and the machine were the defining forces in the age. The heroes of the exhibition were pragmatic men (and women), comfortable with machines, who tinkered, made adjustments, tried out new techniques. Whether self-made, or professionally trained, the engineers and technicians who made the machines at the Centennial were masters of new skills and advanced knowledge. (Mark Twain's *A Connecticut Yankee in King Arthur's Court*, 1889, had such a man, an ingenious Yankee mechanic, as hero, but the application of gunpowder, electricity and 19th-century industrial methods was not a success. Despite the arrival of 500 knights in armour on bicycles, the tale comes to a pessimistic conclusion. Twain had little faith

Thomas Alva Edison (1847–1931) was the greatest of all American inventors. He first demonstrated the phonograph in 1877 (right) which produced sounds recorded as grooves cut into wax cylinders. The cylinder was rotated by the hand crank. Edison expected the phonograph to be mainly of use as an office dictating machine, and its full potential for recording music was not realized until the 20th century.

in the ability of human nature to accept and keep pace with scientific knowledge.) The humble origins of these engineers and the stories of their early experiments in barn loft and stable were celebrated in tales for children. Their achievements lit up the skies:

1878 first store (Wanamaker's in Philadelphia) lit by electric light
1879 streets in Cleveland, Ohio, lit by carbon arc lamps
1880 incandescent electric light
1882 first public electric lighting system (New York City)

In the eyes of disillusioned commentators, the problems of democracy were far more intractable than the laws of physics. "I fear the community has lost all moral sense and moral tone," noted George Templeton Strong in his diary in 1872, "and is fast becoming too rotten to live. We are seriously threatened by social disintegration and a general smash." Cities had grown larger and more complex; buildings were taller; the streets dirtier and more crowded. The government of cities remained mired in corruption and inefficiency. The notorious Tweed Ring had been exposed and booted out of

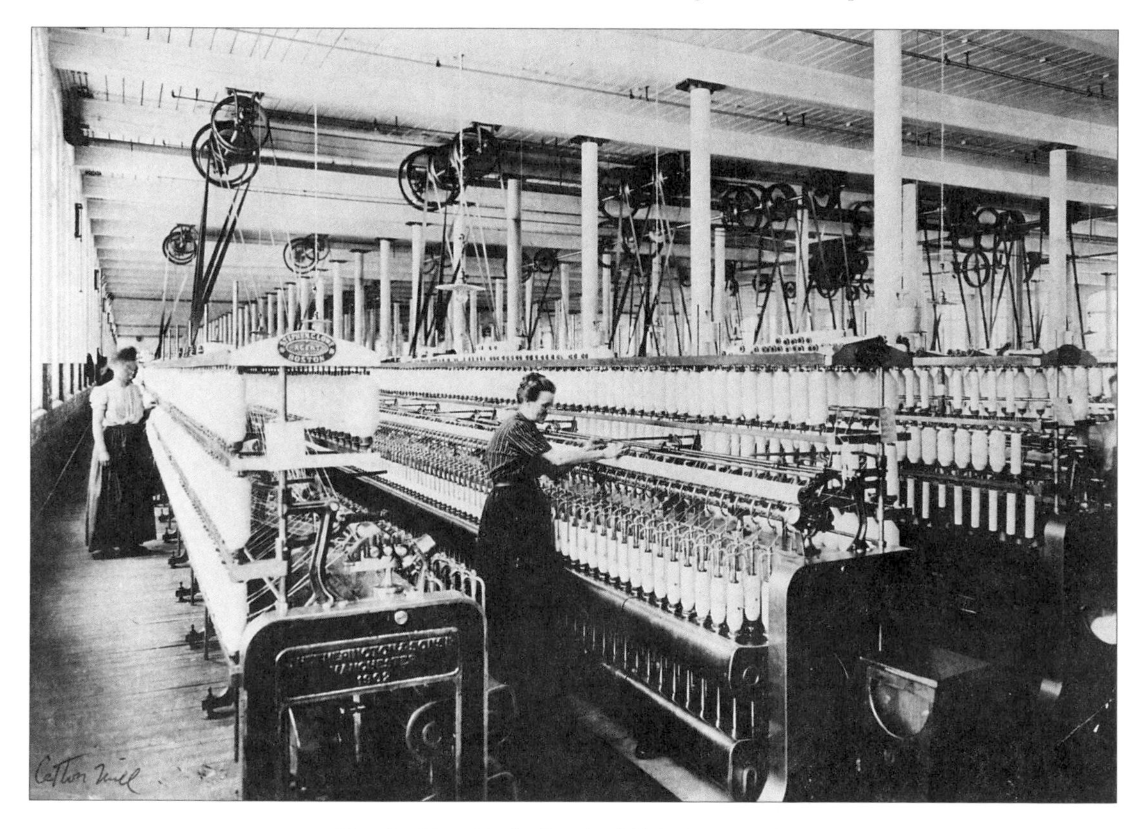

The industrialization of the South in the 20th century was led by the construction of large cotton mills. Young farm-raised women in New England in the 1830s found work in the textile mills, and their Southern counterparts found industrial work in the cotton mills. The machines were manufactured in New England (Manchester, New Hampshire), and sold by Boston agents for installation in Southern plants. Cotton thread would then be shipped to factories where cloth was produced, and then sent to New York where garments were cut, assembled and then sold to Southern stores.

office in New York City in 1871, but smaller Rings prospered in virtually every city and at every level of government across the nation. Individuals of wealth and social prominence regarded public life with disdain, leaving public affairs to the avaricious.

When the Brooklyn Bridge was opened on 24 May 1883, the man chosen to present the principal address was Abram S. Hewitt, whose iron works in New Jersey had perfected the manufacture of wrought iron beams, had given the Bessemer process its first trial at his Phillipsburg works, and who had introduced the Siemens-Martin open hearth process into the United States. Hewitt was a member of Congress, and had been the national chairman of

Newspapers in the mid-19th century were set by hand, and printed by steam-powered presses. The mechanical linotype, and the electric press (the International Printing Press, right), greatly expanded the capacity of the industry to reach ever larger numbers of readers. It was an example of technological change strengthening the democratic foundations of American society.

the Democratic Party. His address opening the Brooklyn Bridge sought to apply the scientific and engineering lessons of the bridge to the management of the city. He argued that the city should "imitate the example of the [Tweed] Ring and organize the intelligence of the community for its government."

> "A city is made up of infinite interests. They vary from hour to hour, and conflict is the law of their being. Many of the elements of social life are what mathematicians term 'variables of the independent order'. The problem is, to reconcile those conflicting interests and variable elements into one organization which shall work without jar and allow each citizen to pursue his calling, if it be an honest one, in peace and quiet."

What made achievements like the Brooklyn Bridge possible was "organized intelligence" and that is what Hewitt offered in his campaign for mayor in 1886. His victory over the political economist Henry George and the youthful Theodore Roosevelt (all three candidates in 1886 were powerful reformers), was a brief, flickering victory for reform in the political life of the nation.

After visiting the Great Paris Exposition in 1900, the American historian Henry Adams reflected in *The Education of Henry Adams* (1906) upon the meaning of the great hall of dynamos. The small experimental machines on show in Philadelphia in 1876 had become perfected, efficient 40-foot objects, moving at incalculable speed, producing an electric force which an elderly gentleman, born on Beacon Hill in Boston in 1838, might not understand at all. There was little more to do than to pray before "silent and infinite force" as a man might have prayed before a statue of the Virgin a thousand years before. The mechanism of the dynamo would forever remain a mystery to Adams. But its meaning symbolized a "break of continuity", the "sudden irruption of force totally new". He had entered a new age of electricity, atoms and X-rays; in this "new universe" there was "no common scale of measurement" shared by the new age and the old. He concluded, in a dozen volumes of American history, that his attempt to grasp the "necessary sequence of human movement" had failed. "Satisfied that the sequence of man led to nothing," Adams wrote,

> "... and that the sequence of their society could lead no further, while the mere sequence of time was artificial, and the sequence of thought was chaos, he turned at last to the sequence of force ..."

When Americans celebrated the new century, it was the "sequence of force" which they welcomed. No one could foresee how far force was to take them.

Charles Lindbergh's solo non-stop flight across the Atlantic in the "Spirit of St Louis" (seen below), *in a field near Paris) in 1927 enchanted the American people. The handsome, boyish aviator seemed to embody the mastery of machines and of the natural world which the American people now believed were within their grasp.*

Civil War 1861–1865

The Civil War which began at Fort Sumter in 1861 changed American society more profoundly than any other military conflict.

"It's all my fault ... it's all my fault." General Robert E. Lee, on Pickett's charge, Gettysburg, 3 July 1863

The presidential election of 1856 brought the South decisively behind the Democratic Party. Four years later, the "solid North" voted for Lincoln and the Republican Party. The traditional custodians of compromise, the Whig Party, had tried to hold on to their supporters in the north and south—and disintegrated. With the election of Lincoln, the divisions about slavery, states' rights and territorial expansion were locked in place. Southerners believed that the millions of dollars they had invested in slaves was in jeopardy if they remained within the Union. No matter how sincerely Northern politicians reassured the South, conviction grew that economic and political ruination would follow the election of a Republican president in 1860.

The war which began on April 12, 1861, with the cannon fire on Fort Sumter, was the most conservative in American history. On both sides, the struggle to preserve the old order ultimately caused the old order's destruction. Slavery was ended; the cause of states' rights was swept away as the federal government set about imposing a national system of economic regulations. African-Americans were admitted to serve in the US Army for the first time since 1792. (Segregated black units persisted in the military until after World War II). The traditional Southern codes of honour melted away with desertion and military defeat. Women came to play an important role in military medicine and the US Sanitary Commission. On all sides, the Civil War was a surprise. There was no slave uprising in the South, and even during General William T. Sherman's "march to the sea" (Atlanta to Savannah, 1864), rape of Southern women was unknown.

It was, above all, a war of unprecedented blood-letting. The North, with a population of 21 million, confronted a numerically smaller Confederacy (9 million, 3.5 million of which were slaves). The North could (and did) lose battle after battle, and—if its nerve held—still win the war. The loss of 175,000 men in the first 27 months of the war left the Confederacy bleeding to death. The total Southern dead of 260,000 was far exceeded by 360,000 Northern dead, but the proportion of Southern losses was much greater. In relation to the US population of about 30 million, these were heavier losses than Britain sustained in World War I.

Gettysburg, 1863: the battle scene, painted by Thure de Thrulstrop, captures the moment when the outcome was not yet decided. The failure of Pickett's charge and the defeat of Lee by the Union Army under Meade ended any Southern hopes of administering a demoralizing blow to Northern public opinion.

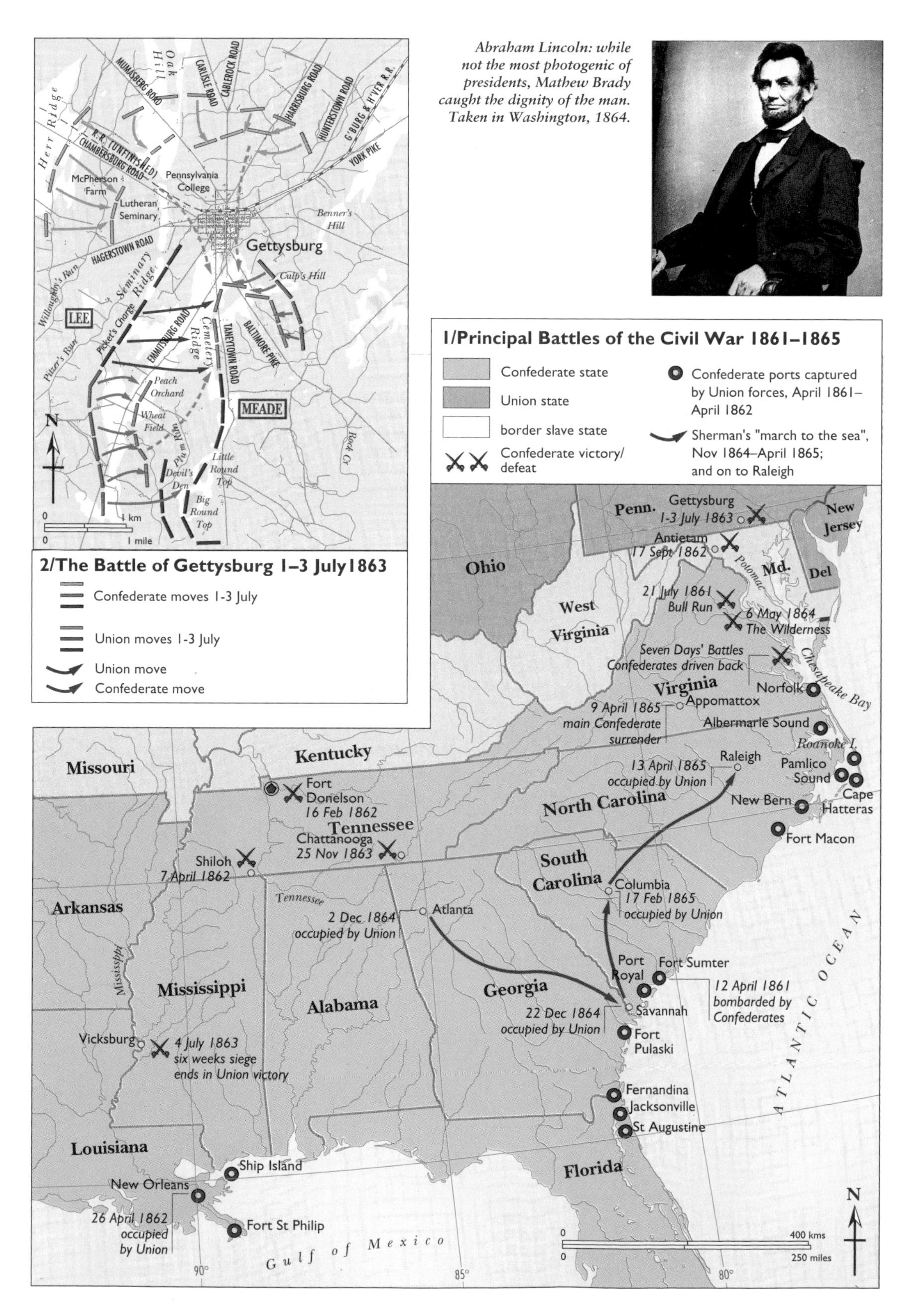

Abraham Lincoln: while not the most photogenic of presidents, Mathew Brady caught the dignity of the man. Taken in Washington, 1864.

British North-America Act 1867

"Responsible" government, and then confederation, began the process of uniting the provinces of British North America.

"I expected to find a contest between a government and a people: I found two nations warring in the bosom of a single state."
Lord Durham's *Report of the Affairs of British North America*, 1839, proposing legislative union of the two Canadas

In 1791 the majority of the 250,000 inhabitants of British North America were of French origin. By 1845, Canada was by a substantial majority "British". Over half a million immigrants, mainly from Wales, Scotland and Ireland, had tipped the balance of the colony's ethnic composition.

Politics remained the preserve of a small élite of colonial civil servants, merchants and large landowners. They were the people who lived in houses of stone and brick. The others, the small farmers and trappers in log huts, struggled to gain a subsistence living from the land. The majority faced a life of exhausting labour and isolation. While their husbands fished in the summer, or disappeared into the woods in pursuit of fur-bearing animals, like the beaver, Canadian women raised their families and worked on the farm. (On average, women in French Canada had seven children.) The poorest and least successful were likely to be the most mobile: it was a society of people constantly moving in search of betterment, a society of people with weak local roots.

"Responsible" government—long the cause of Liberal reformers—was achieved by Nova Scotia and the Province of Canada (Upper and Lower, then renamed Canada East and West) in 1848, with the other provinces following in the 1850s. The breakdown in 1861 of the union in the United States did not discourage discussion—led by John A. Macdonald, the Scots-born leader of the Conservatives—about federal union in Canada.

The "Dominion of Canada" was created by the North-America Act, passed in the British Parliament in 1867, assuring the French a province in which they were dominant (Québec). Two of the Maritime Provinces joined (Nova Scotia and

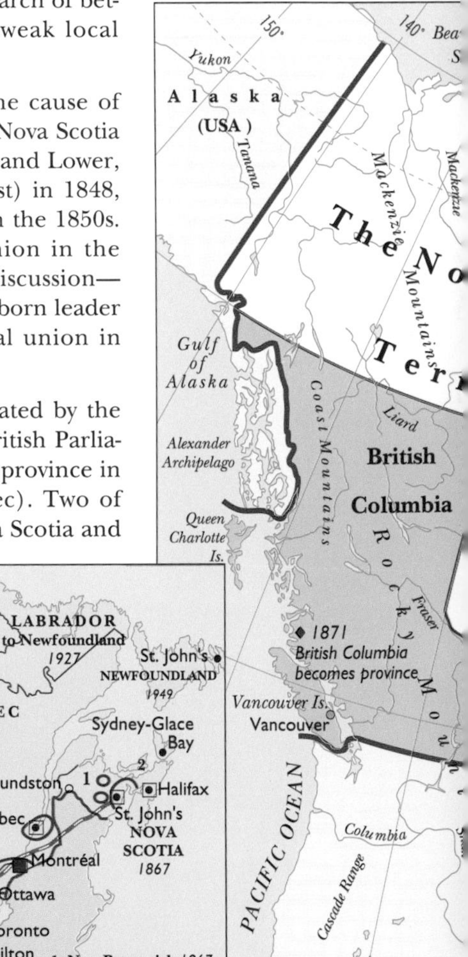

1/The development of Canada, 1867–1920

population 1871:
- □ town of 25,000 to 100,000 people
- ■ town of over 100,000 people

population 1911:
- • town of 25,000 to 100,000 people
- • town of over 100,000 people
- ○ other town
- ▬ main industrial regions
- *1871* entry into Dominion

1 New Brunswick *1867*
2 Prince Edward Island *1873*

New Brunswick), but Newfoundland voted decisively against union in 1869. (It remained outside the confederation until 1949.) Generous financial terms brought Manitoba, British Columbia (both 1870) and Prince Edward Island (1873) into the union. The opening of the Canadian Pacific Railway to British Columbia in 1885 was the symbolic completion of confederation.

Delegates from Canada, Nova Scotia and New Brunswick met in London to decide the basis for union. The United States government pressured the Colonial Office to block the name favoured by the Canadian conservatives, "Kingdom of Canada". "Dominion" was an acceptable compromise, and was enshrined in the British North -America Act which was signed by Queen Victoria on 29 March 1867.

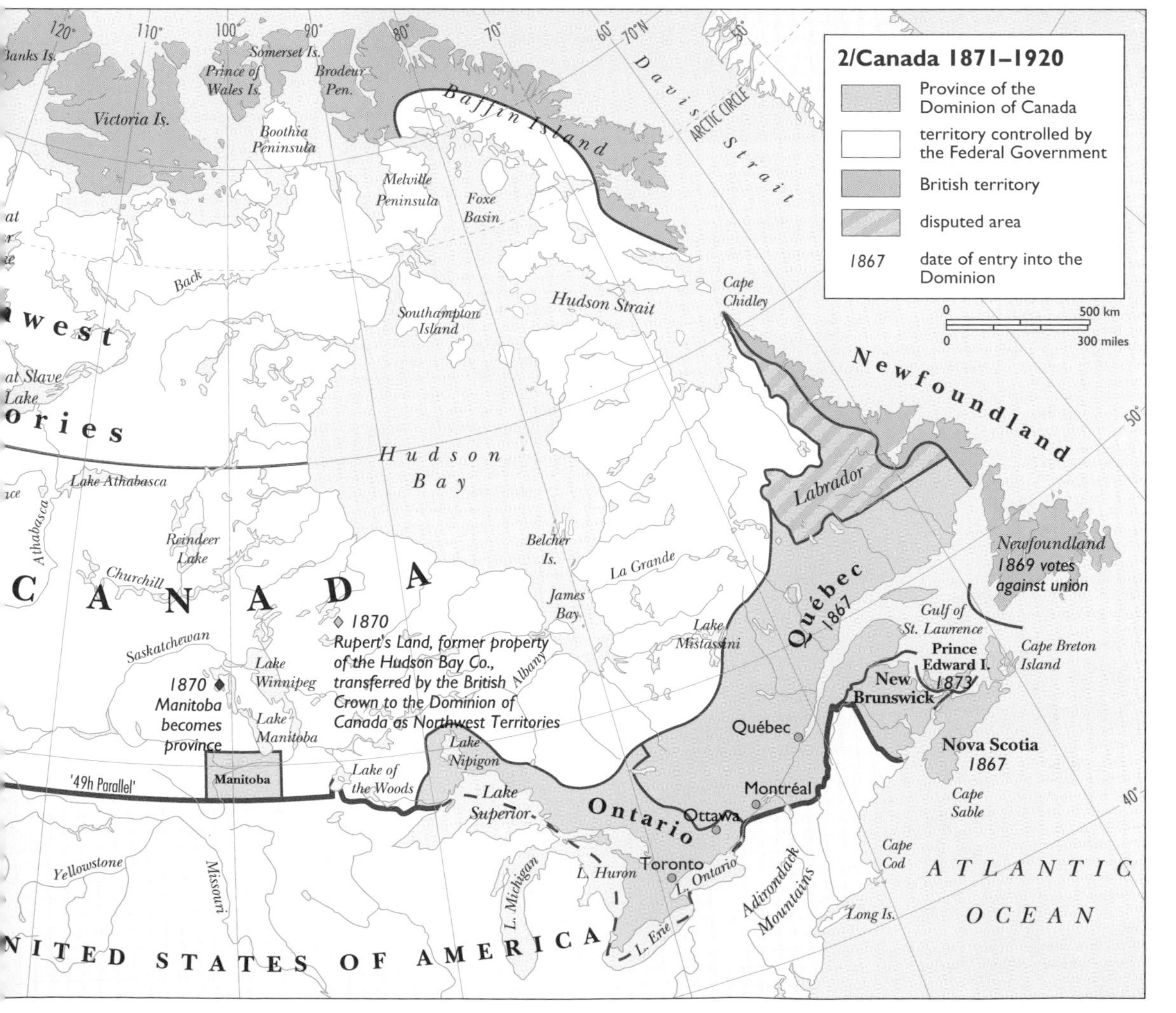

Alaska – "Seward's Folly"

The Russians came to Alaska to hunt the sea-otter, but sold their claim to the United States in 1867.

"... on a good year there is bad weather for two-thirds of the time and only one-third of days are clear and moderate."
Kyril T. Khlebnikov on the trading port of New Arkhangel, Sitka Island in 1817

The Russian claim to Alaska was based upon the voyages of discovery of Vitus Bering, who confirmed that Asia and North America were separate continents. He reached the Alaskan mainland in 1741. The first Russian presence came in 1784, when a trading post was established on Kodiak Island.

The Russian-American Fur Company, granted a monopoly of Alaskan trade, founded a permanent settlement on Sitka Island in 1799. The native inhabitants, the Tlingits, opposed their presence and forced the Russians to build a fortified stockade, New Arkhangel, in 1804. Twelve years later there were 620 residents, 435 males and 185 females. Russian dignitaries donated a learned library to the settlement.

Trade was highly prosperous (profits of 250 per cent were made on pelts sold at San Francisco and Monterey). In 1810 John Jacob Astor, the New York fur trader, negotiated an agreement to sell Russian furs in Canton for a 5 per cent commission.

The Russians helped to dominate the fur trade by extending their control southward along the coast (Fort Ross, 1812). In 1815 a mission was sent to Hawaii to see if annexation was possible, but the mission failed. The Russian claim to territory as far south as 51°N brought them into conflict with the Spanish as well as the Americans, who were disputing the Oregon Territory with Britain. The Russians became alarmed at the increased presence of American trading vessels, and were prepared to resist the commercial interest in the north of the Hudson Bay Company. (Fort Yukon, at the junction of the Yukon and Porcupine Rivers, was founded in 1847). Preoccupation with military affairs in the west (Crimean War, 1854–56) made Alaska less central to Russian interests. When Secretary of State William H. Seward made an attractive offer of $7.2m in 1867, the Russians accepted. Critics scorned "Seward's Folly", but the salmon trade alone repaid the cost of the purchase. Alaska achieved full territorial status in 1912, and was granted statehood in 1959.

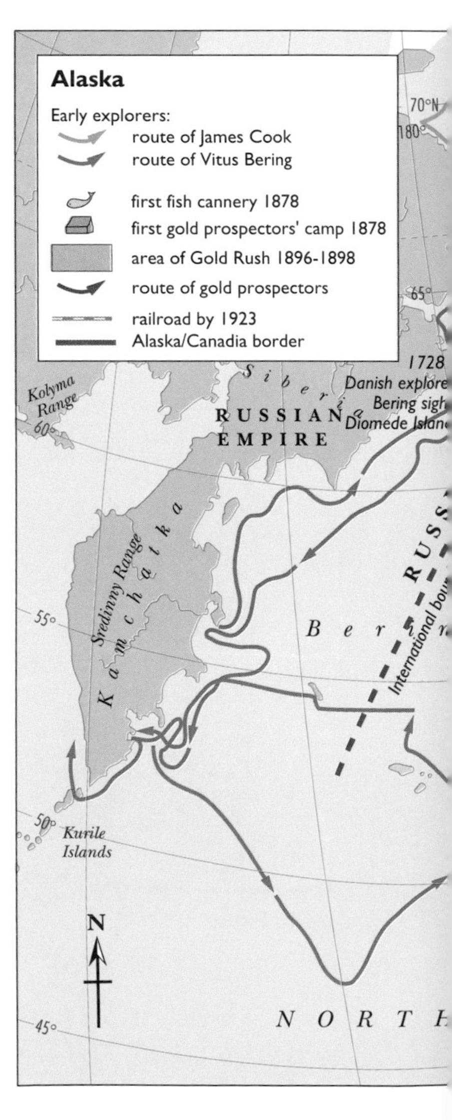

There was considerable interest in the scientific circles of St. Petersburg in the tribal peoples who inhabited the Alaskan islands. An Aleut woman, with her baskets, nets and bone tools, was sketched by Levashev at Unalaska Island, 1767.

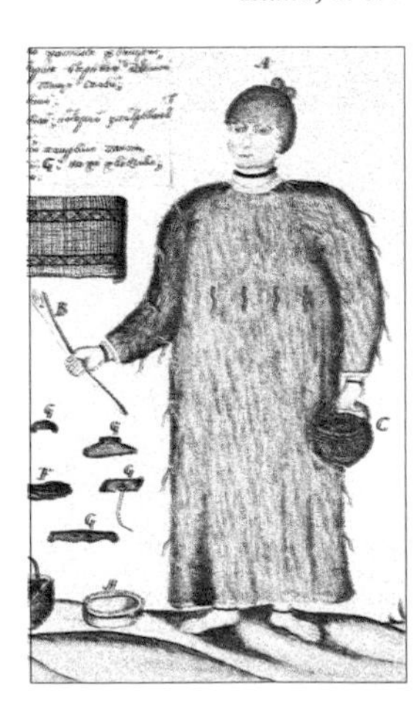

Captain James Cook's Resolution *and* Discovery *stopped at Nootka Bay on the unsuccessful expedition in 1778 to find a northern passage to the Atlantic. The expedition artist, John Webber, drew the canoes of the Nootka Indians coming out to trade with the visitors.*

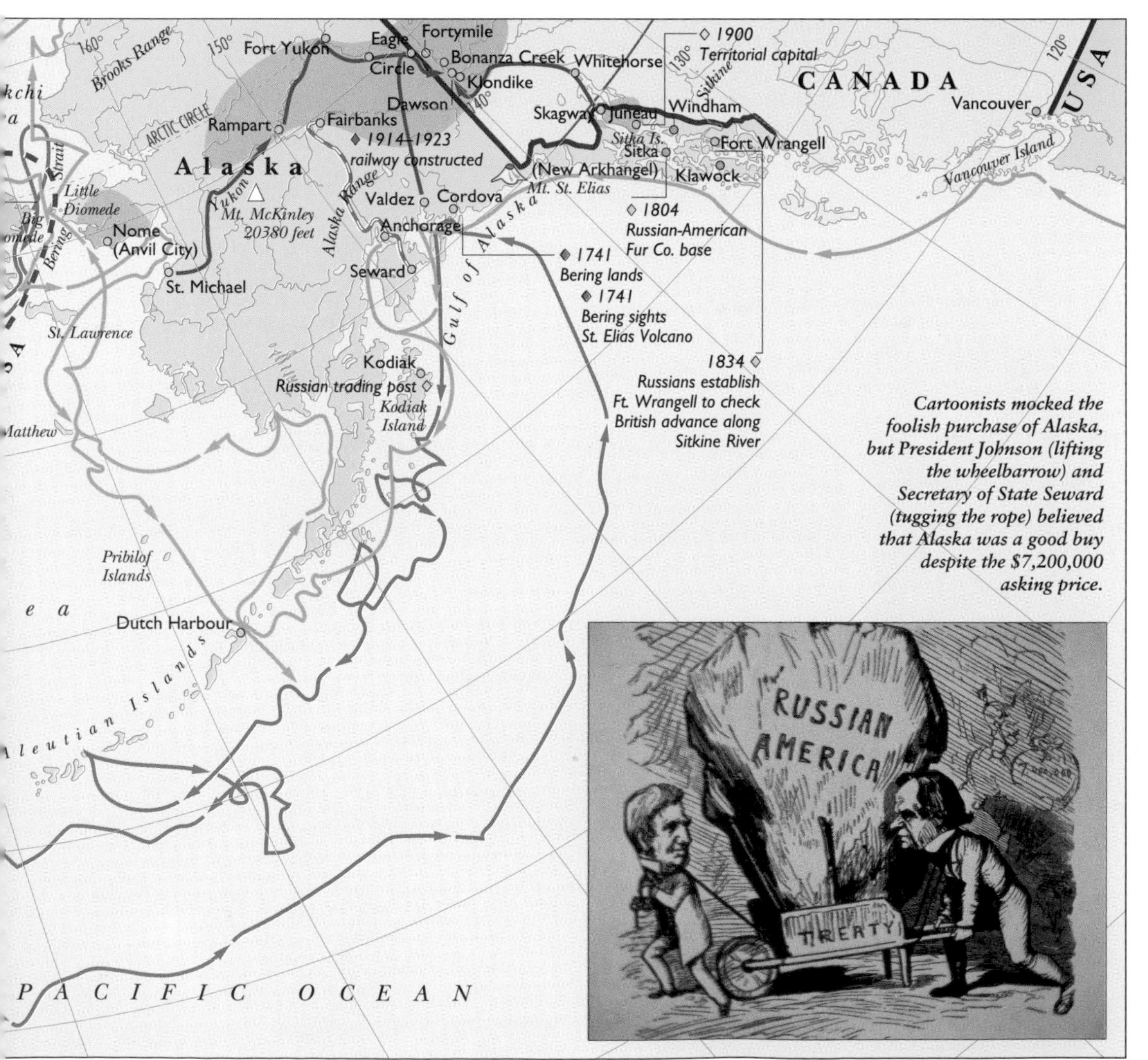

Cartoonists mocked the foolish purchase of Alaska, but President Johnson (lifting the wheelbarrow) and Secretary of State Seward (tugging the rope) believed that Alaska was a good buy despite the $7,200,000 asking price.

The Age of Chicago

Chicago's place at the centre of the national rail network, and its pioneering of the skyscraper, made it the great symbol of modernity in American life.

> *"A vast flat city, cut to bits by tracks of steel."*
> Waldo Frank on Chicago in *Our America*, 1919

On 10 May 1869, the construction gangs on the Central Pacific railway line met the Union Pacific line at Promontory Point, near Ogden, Utah. A gold spike was tapped into the track to symbolize the completion of the first trans-continental railway line. News of the completion was telegraphed across the nation. Nowhere were the celebrations more enthusiastic than in Chicago. The finished line ran from Omaha, Nebraska, to Sacramento, California. The main rail link with Omaha ran to Chicago. Until the completion in 1881 of the line which connected Kansas City to California, Chicago virtually monopolized traffic across the nation.

The great cattle drives from Texas to Abilene, Kansas, brought tens of thousands of head of cattle to the slaughterhouses of Chicago, from where they were shipped in refrigerated railway cars to the cities on the east coast. The national railway network grew from 53,000 miles to 94,000 in the 1870s, and to 167,000 miles by 1890. With each new line, Chicago manufacturers looked forward to new markets, new opportunities. Quick to seize the opportunities on offer, Chicago was rebuilt after the great fire of 1871 with tall office buildings. In 1885, the world's first steel frame construction began with William LeBaron Jenney's Home Life Insurance building in Chicago. The electric elevator, first demonstrated by Elisha Graves Otis in 1852, and Alexander Graham Bell's telephone (demonstrated in 1876), made possible the skyscrapers which symbolized Chicago, and the modern urban world.

Chicago was also the great battlefield of social change in industrial America. The four anarchists who were hanged for the explosion in Haymarket Square in 1886, and the Pullman railway strike of 1894, nearly destroyed the 19th-century labour movement, but social concern—represented by Upton Sinclair's novel, *The Jungle* (1906)—continued to find Chicago the most ominous of great American cities, where the power of the rich ran unchecked.

The opening of the Union Pacific line from Omaha made possible direct rail travel between Chicago and San Francisco. This poster advertises the opening, 1869.

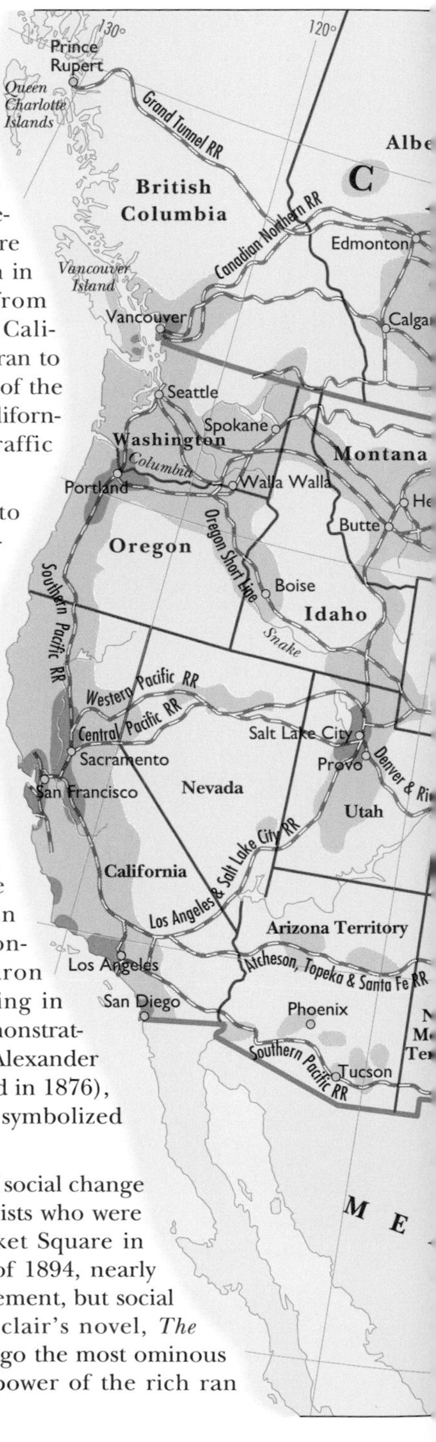

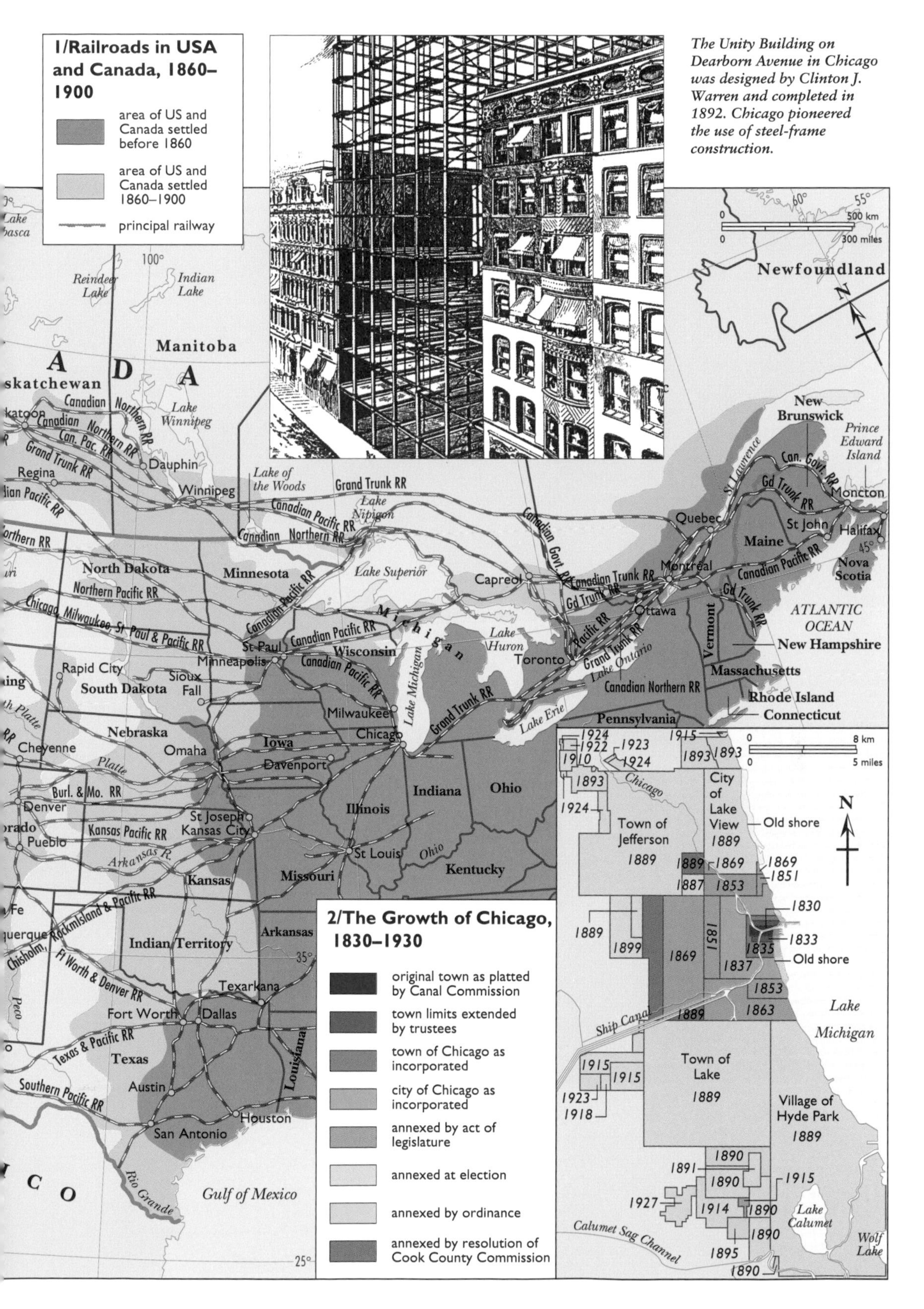

The Unity Building on Dearborn Avenue in Chicago was designed by Clinton J. Warren and completed in 1892. Chicago pioneered the use of steel-frame construction.

Frontier

The "Winning of the West" defined American values and popular mythology in the decades after the end of the Civil War.

"If California lies beyond these mountains [the Sierra Nevada] we shall never be able to reach it."
John Bidwell, member of the first emigrant party to cross overland to California in 1841

Americans regarded the lands beyond the settled territories as a wilderness, and then a "frontier". When it was announced in the 1890 census that the frontier was closed, that there was no longer a continuous line of unsettled land, there was much discussion about the significance of this moment in the nation's history. America without a frontier was an unfamiliar place, where cities and factories—and not open land—would shape the national character.

The frontier passed so suddenly, so irrevocably, that the imagination of the nation has perhaps never quite recovered from the trauma of its loss. The topography of the land to the west of the Mississippi River was daunting, with its deserts, badlands, arid plains, scorching summer heat (a high of 134°F in Death Valley) and frozen winters (as low as –60°F). Great gorges like the Grand Canyon dissected the plateau. Beyond the plains lay the Rocky Mountains, reaching down from the Canadian border. In southern Colorado there are 52 peaks above 14,000 feet. The topography remained, but little else was permanent. The great era of exploration (begun by Lewis and Clark, 1804–06) and fur trappers (1820s–1830s) ended when emigrants entered the Great Plains in Conestoga wagons in 1834. The completion of the transcontinental railway in 1869 ended the era of the wagon train. The Pony Express, running nearly 2000 miles from Saint

American painters found a ready market for western landscapes, especially those which combined patriotic and romantic themes. A lithograph dated c.1860.

Joseph, Missouri, to Sacramento, California, only functioned from 1860 to 1861. The cattle trails, along which 10 million longhorns walked from Texas to the railways in Kansas and Missouri, survived 20 years from 1866.

There were 75 million buffalo roaming the Plains at mid-century. For generations an Indian way of life had been based on the horse and buffalo hunt. Hunters wiped out the buffalo in the 1870s, and the Indians were defeated by the U S Cavalry in the following decade. The way was now clear for the settlement of the West. What had been an unknown space was mapped, surveyed and settled in a single generation.

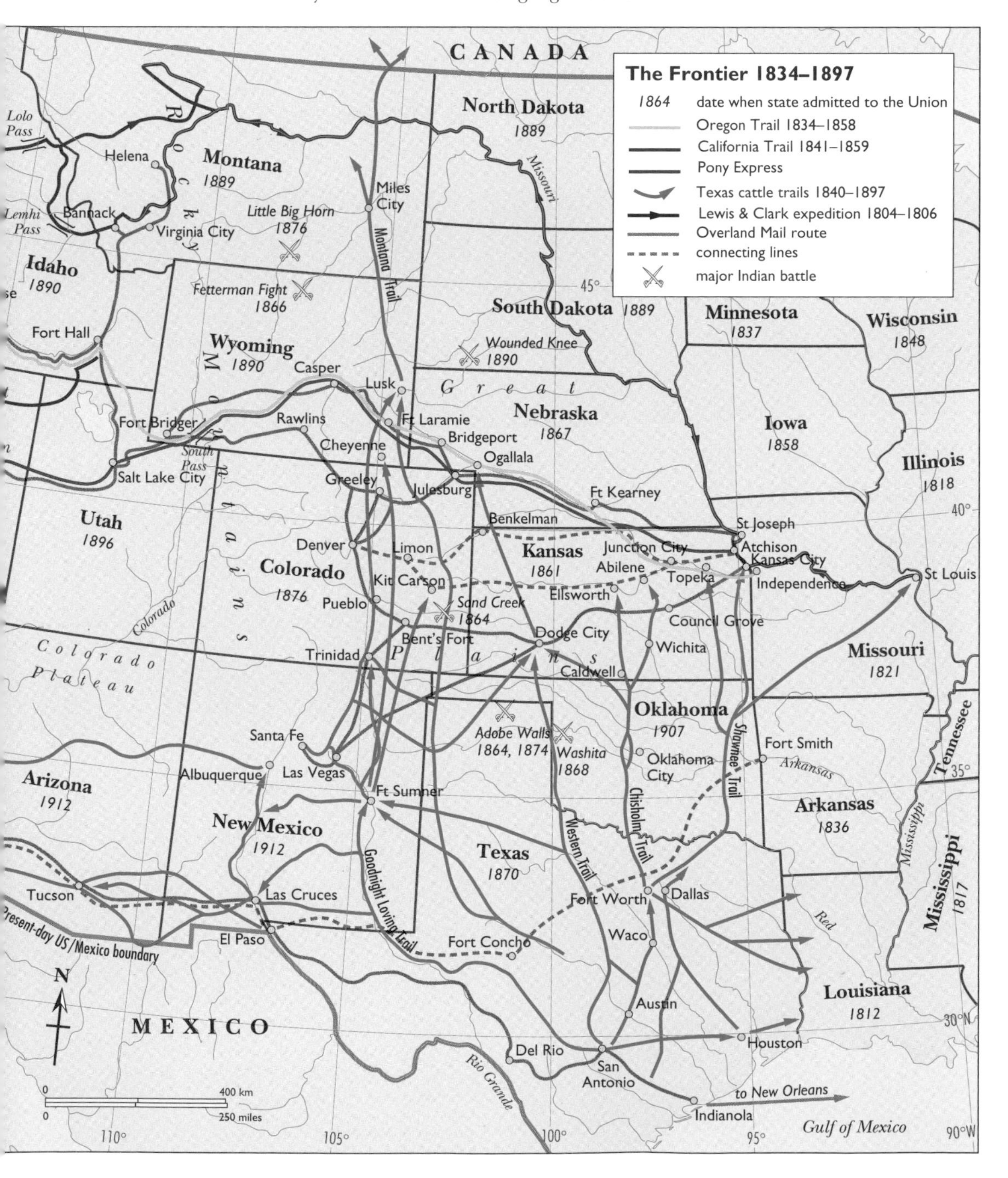

An Imperial Power

The Cuban struggle for independence from Spain initiated the era of sustained American intervention in the Caribbean, Central America and the Pacific.

"Our place must be great among nations." President Theodore Roosevelt addressing Congress, 1902

The American people were by persuasion isolationist; memories of the struggle against the greatest imperial power, Britain, inclined public opinion to oppose imperial ventures. After the war in Mexico in 1846-8, the US Army returned across the Rio Grande. Territorial expansion—and the doctrine of "Manifest Destiny"—commanded widespread support, but not to the point of conquest and oppression.

Yet by the presidential election of 1900, the public wholeheartedly supported McKinley and an aggressively expansionist national policy. Despite the fears of anti-imperialists that the acquisitions of an empire would endanger

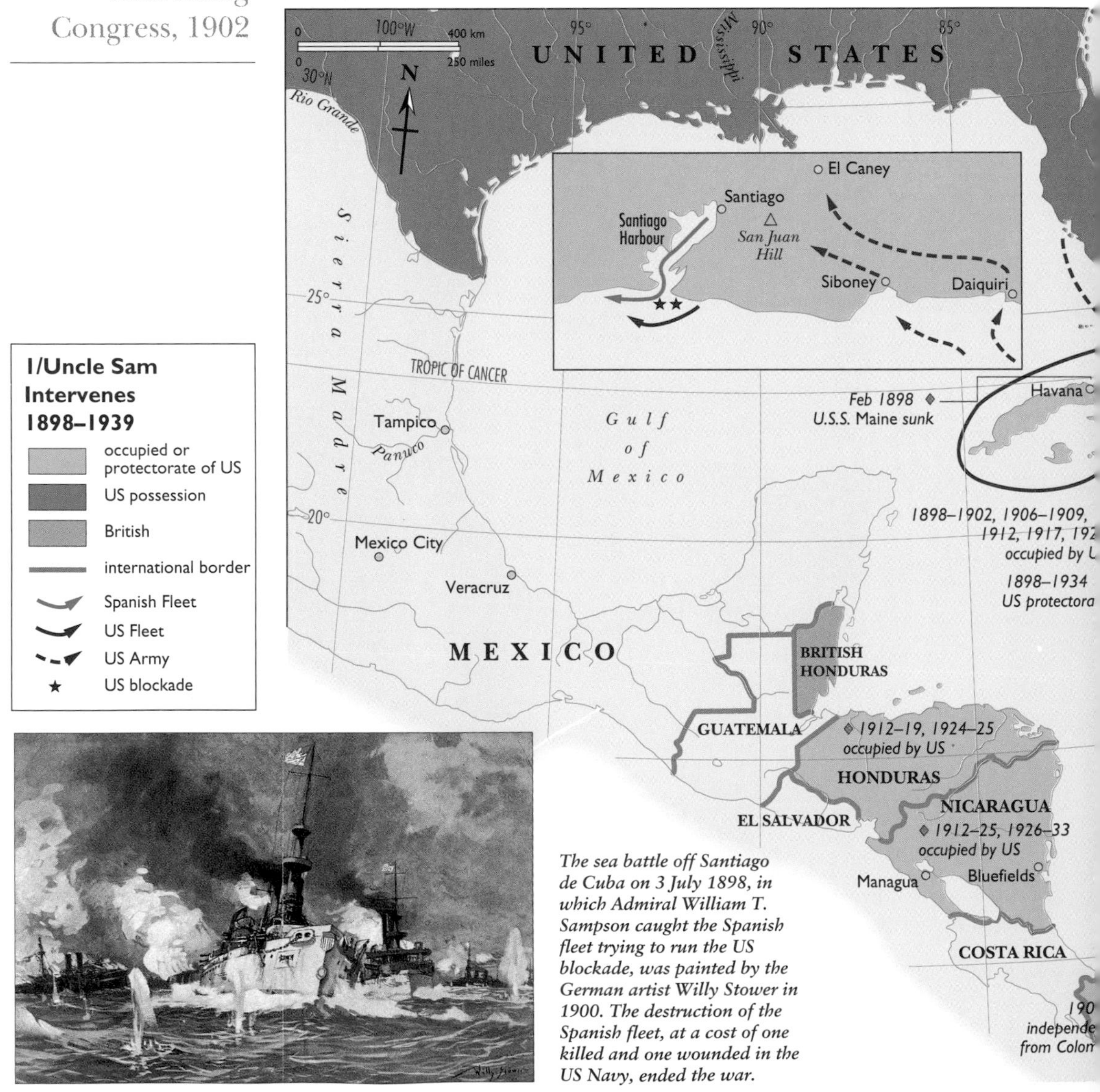

The sea battle off Santiago de Cuba on 3 July 1898, in which Admiral William T. Sampson caught the Spanish fleet trying to run the US blockade, was painted by the German artist Willy Stower in 1900. The destruction of the Spanish fleet, at a cost of one killed and one wounded in the US Navy, ended the war.

American liberty and the republican form of government, the imperialists—whose natural home was the Republican Party—argued that survival depended upon expansion. Nations, no less than species, were engaged in a struggle in which only the fittest would survive. Those societies which fell behind in the race for markets would be deprived of the means of sustenance by more powerful peoples. Falling prey to internal conflict, they would grow weaker. Only those who expanded would grow wealthy, numerous and powerful. External expansion was portrayed as the sole guarantor of commercial prosperity and the best way to secure domestic tranquillity. "We must have the [Chinese] market," argued a senator, "or we shall have revolution."

The Cuban struggle for independence from Spain gave the American expansionists the green light, but the decision to attack the Spanish fleet in Manila Bay revealed the broader purpose of the intervention. For virtually the rest of the century, the United States government has treated the Caribbean and Central American states as a "sphere of interest" which has resulted in military occupations, annexations, and the maintenance of periods of "protectorate" status over Panama, Haiti and the Dominican Republic and Cuba. The interests of American capital have been well-protected.

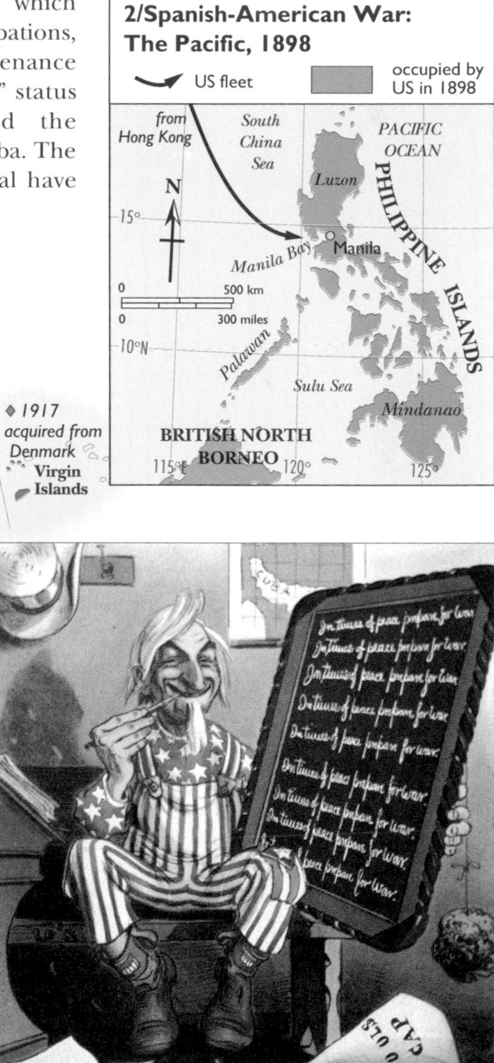

The unpreparedness of the army was the occasion of Grant Hamilton's cartoon of Uncle Sam as a dilatory schoolboy assigned to write out what he had learned from the war with Spain: "In times of peace prepare for war."

Skyscraper

The skyscraper gave symbolic expression to the power and ambition of American commercial life.

"By their mass and proportion [skyscrapers] convey in some large elemental sense an idea of [the] great, stable, conserving forces of modern civilization."
John Wellborn Root, architect, June 1890

After the great fire of 1871, the centre of Chicago, a metropolis of 300,000 residents, was a charred ruin. The city's grid plan and railway network restricted the amount of space available for city centre offices; the cost of land rose sharply, and developers instructed their engineers and architects to build the tallest structures possible.

Richard Morris Hunt's *New York Tribune* building in Manhattan (1874) was nine storeys high. It was one of the first commercial buildings to install an elevator. Daniel Burnham and John Wellborn Root's Montauk Building (1881–1882) was the first 10-storey structure in Chicago, built using a new principal of construction: a cast-iron frame which bore the entire weight of the structure.

Tall buildings created structural problems in the sandy Chicago soil, and a "floating raft" foundation—consisting of a slab of concrete 20 feet deep reinforced with steel rails—was needed for the tallest structures. The idea of "Chicago construction", the "floating raft", metal cage, and austere style of the skyscrapers was extended by William LeBaron Jenney's Home Insurance Building (1883–1884) and Holabird and Roche's Tacoma Building (1887–1888).

The metal cage made taller buildings possible—doubling or tripling the space available for rental—and had a dramatic impact upon their appearance. Non-loadbearing walls could be thinner and more decorative. Larger windows were possible. Increased interior light, even after the invention of the incandescent bulb by Edison in 1879, meant better working conditions, increased productivity—and higher rents.

The tall office building was a new thing in the history of architecture; there was no language or style which architects could naturally apply to their pro-

A view of midtown Manhattan, taken from a helicopter. The memorable art deco pinnacle of the Chrysler Building is in the centre. The (former) Pan Am Tower, behind the Chrysler, marks the beginning of Park Avenue. Central Park is visible on the right. In the distance are the Hudson River and New Jersey.

jects. Louis Sullivan's celebrated article, "The Tall Office Building Artistically Considered" (1896), argued passionately against the imposition of historical styles for a uniquely modern and functional structure.

New buildings in New York—Ernest Flagg's Singer Building (600 feet, 1908) and Cass Gilbert's Woolworth Building (792 feet, 1913)—were as ornate as Gothic cathedrals. By 1914 it was assumed that the design for a skyscraper should contain some historical reference. The spirit of the Chicago skyscraper was reborn in the 1920s, when Mies van der Rohe proposed a skyscraper sheathed entirely in glass. But the *Chicago Tribune* architectural competition in 1922 was won by a Gothic tower designed by Howells and Hood. Flying buttresses, and not the autonomous, pure tower of Eliel Saarinen (which won second place), remained the preferred style into the next decade.

Left: *A postcard view of the Woolworth Building, designed by Cass Gilbert, which was completed in 1913. It was built for Frank Woolworth, who instructed the architect: "I do not want a building. I want something that will be an adornment to the city."*

Right: *The Times Tower, designed by Eidlitz and MacKenzie in 1904, occupies a wedge-shaped plot created where Broadway crosses Seventh Avenue at West 42nd Street. In 1928 a great bank of 14,800 lights were installed on the four sides of the Times Tower to create the world's first moving electrical sign. The* New York Times *once occupied this building. In the course of remodelling in 1966, the terracotta Italian skin was stripped off and replaced with white marble.*

V: An Urban People

The new sciences of "race", seeking to define the superiority of the Anglo-Saxon people, attempted to halt the flood of immigrants into urban America.

The gap which Abram S. Hewitt noted in 1883 between the "organised intelligence" of science, technology and reason, and that of the crooks and political bosses who ran the American cities, had grown into a Grand Canyon by the turn of the century. At heart, Hewitt was a man of the Enlightenment, who believed that the application of reason would produce a stream of material benefits and improvements for society. His successful ironworks in New Jersey were a demonstration of such benefits. So too had been Cyrus McCormick, who in 1831 perfected the mechanical reaper which bore his name. The flight of the Wright brothers at Kitty Hawk, North Carolina, on 17 December 1903 was a similar triumph. They were proprietors of a bicycle shop whose curiosity and tinkering opened the door to the wild notion of powered flight.

"... we want people who can understand us, accept our ways, and become whole-hearted partners in the great task of perfecting OUR America. But we most emphatically do not want people, NO MATTER HOW INDUSTRIOUS OR HOW INTELLIGENT, who don't like us, don't fit into the national fabric, and instinctively want to change our America into something different, based upon different ideas and attitudes toward life."

Lothrop Stoddard

Orville (1871–1948) and Wilbur Wright (1867–1912), who were inspired by the work of Otto Lilienthal with gliders, worked with piloted gliders before building their first powered machine in 1903.

Opposed to such triumphs of practical reason were the dark forces of vice, greed and corruption. Between the "sunlight" world of science and the "gaslight" realm of saloons, gambling dens and brothels there was unremitting warfare. Everywhere there were signs of advances made possible by science: material changes, based on science and technology, were transforming the ways people lived and the way they earned a living. At the same time the new scientific ideas were disturbing the most deeply-rooted assumptions about human nature, community and the social order.

Attempts to apply the principles of science to society, and to the difficult issues surrounding the growing inequalities of wealth, produced policies which were cloaked in rationality, but which were far from scientific. The

The slow-moving traffic of horse-drawn wagons and stages in 19th-century cities encouraged transportation companies to seek alternatives. The laying of rails in city streets for horse-drawn omnibuses, the use of steam-engines, the construction of elevated railways, and then the digging of tunnels for subways, each promised an escape from the endless ill-tempered, slow-moving procession of city streets. The fast-moving automobile seemed in its early years to offer a perfect liberation from crowded streets. As cars became more popular in the 1920s, they in turn were responsible for even more congested city transportation. Columbus Circle in New York (right, *in the 1920s) was the scene of a daily traffic jam, as buses, trucks and passenger cars sought to head downtown from 59th Street.*

attempt by the English writer Herbert Spencer in *The Study of Sociology* (1872–73) to apply Darwin's general law of evolution to society, was read in America as a powerful revelation of social law. Spencer believed that there was a science of society which could chart the course of social evolution. Altering or opposing the laws of evolution would derail progress. Society could do nothing better than let development follow its natural course. (Thus the French phrase *laissez-faire* won acceptance in the English-speaking world for government policy that in principle did not interfere in social or economic processes.) The Spencerians, or "Social Darwinists", repudiated state interference. They were hostile to state aid to the poor (though not to voluntary charity). If high mortality rates and illness indicated that "Nature" was seeking to do away with the tenement dweller, how could society have the temerity to interfere? Humans must adapt, and those who adapted most successfully were the most likely to survive. Social laws were based upon the necessity of the struggle (the process of "natural selection"), and the survival of the fittest.

In Sunday Schools and in newspaper editorials, in high-brow magazines and popular scientific journals, disciples of Spencer taught Americans that government interference was a threat to the gospel of progress. John D. Rockefeller, founder of the Standard Oil Company, explained that the growth of a large business was "merely a survival of the fittest". Andrew Carnegie, whose steel company dominated Pittsburgh, explained that though the law of competition was sometimes "hard on the individual, it is best for the race". William Graham Sumner of Yale University argued that struggle, competition and inequality were fundamental laws of nature. The notion that "we ought to see to it that everyone has an existence worthy of a

The immigrant family became one of the great icons of early 20th-century America. Whether as a source of fear because they would dilute Anglo-Saxon racial purity, or as a threat to law and order, or because they were prey for the crooked bosses who ruled the city, the immigrant was frequently represented negatively. The photographer of the family above (Ellis Island, 1911) clearly saw the immigrant family as a source of hope.

human being" was, for Sumner, a highly dangerous "noble sentiment", which would have the effect of reducing men to the condition of paupers. If deprived of the necessity to provide for themselves, character would be ruined and family life corrupted, leaving the recipients of what we would call "welfare" in a state of moral degradation.

Superior peoples

The American victory over Spain in 1898 seemed no less a triumph of natural selection than the vast wealth of Rockefeller and Carnegie was a sign of their superior abilities in the struggle for corporate life. Imperialists, wrapping themselves in Darwinian language, presented militarism and expansion as furthering the needs of the race. The superior Anglo-Saxon (Aryan, in the case of German racial thought) would triumph over the weaker and backward races. Americans were well-schooled in the idea of their racial superiority. The centuries of slavery in North America and the wars with the Indians were indeed proof of the superiority of the Anglo-Saxon race. (The term *English-speaking world* came into use as an alternative to *Anglo-Saxon race* when explicit racial language began to lose favour. It helped to overcome American isolationism and was a more appropriate usage for groups whose ancestry was not British.)

Racism became intertwined with American nationalism, patriotism and "scientific" knowledge. Britons lustily brayed the refrain of James Thomson's "Rule Britannia":

> Rule Britannia, rule the waves,
> Britons never will be slaves.

Americans, too, shared the jingoistic pride of the British in their unique historical destiny. The victory over Spain confirmed that they had joined the British in spreading the English language, traditions and blood across the seas and continents. Political leaders saw in the expansion of the Anglo-Saxon race the great challenge facing the nation. "If we stand idly by," warned Theodore Roosevelt in a lecture on "The Strenuous Life",

> "if we seek merely swollen, slothful ease and ignoble peace, if we shrink from the hard contests where men must win at hazard of their lives and at the risk of all they hold dear, then the bolder and stronger peoples will pass us by, and will win for themselves the domination of the world."

Threats to the position of Anglo-Saxon dominance arose in 1905, when for the first time a "white" European power, Russia, had been defeated by the Japanese. The "Yellow Peril", long associated with violent Californian hostility to Chinese immigration, was projected across the world stage as a great confrontation between a declining, decadent white race and a youthful, brutal Asiatic-Mongolian power in the East. This was the theme of Oswald Spengler's *The Decline of the West,* which had a great American vogue in the 1920s, and of *The Passing of the Great Race* by Madison Grant, published in 1916, which warned that the pioneering Nordic peoples were in danger of being swamped by inferior racial stock. Grant called for the segregation of races to avert impending disaster. The enthusiasm of Tom Buchanan in F. Scott Fitzgerald's *The Great Gatsby* (1925), for the (fictional) book by Goddard, *The Rise of the Colored Empires,* neatly captured Madison Grant's form of American racial hysteria and racism.

Anthropologists and sociologists studying the hierarchy of races found that the Anglo-Saxon or Aryan type was superior, and that the Negro or African occupied the lowest position on the evolutionary ladder. Anatomical and physiological differences, from the weight of the brain to the shape and length of the nose, were taken to be measures of inferiority. The scientific community thus gave a gloss of reason and objectivity to the racism and deepening discrimination which blacks faced in the United States. The first full-length cinema feature in America, D.W. Griffith's *The Birth of a Nation* (1915), was an idealization of the historical myth of the Slave South and a justification of the armed defence of the white race agains the threat of ignorant blacks.

What separated the black African from the Anglo-Saxon American also applied to the distinction between the democratic, hard-working, Protestant peoples of northern Europe, who exemplified the pioneer virtues, and the "inferior" races and "degraded peasantry" of Southern and Eastern Europe, who increasingly filled the slums of urban America. Fears of the impending decline of the nation underlay a racial sense of pessimism and alarm.

The "new" immigration

Absolute numbers of immigrants rose sharply from 2.8 million in the 1870s to 5.2 million in the 1880s. After a decline in the economically-depressed 1890s (3.6 million), the total rose steeply to 8.8 million in the first decade of the new century. The majority of the "old" immigrants were drawn from the British Isles, Germany and Scandinavia. The "new" immigrants from 1880 to 1930 (some 27 million) largely came from southern and Eastern Europe.

The new immigration drastically changed the ethnic composition of America. Where there had been 250,000 Jews in the United States in the 1870s, mostly of German origin, by 1927 there were 4 million, overwhelmingly drawn from Russia and Eastern Europe. The Jewish community in the United States, which showed many signs of achieving a reputable place in commercial life, brought with it the German passion for culture and bourgeois respectability. Their observance was, in the German tradition, "reformed", and they spoke German at home, not Yiddish. Until the resort hotels in Saratoga began to exclude Jews in the 1880s, they were highly acculturated into American life. The Jews who arrived with the new immi-

Baseball was one of the things which helped unify the increasingly diverse urban population. The handlebar moustaches of the Brooklyn Dodgers were among the most impressive in the National League.

gration were, in the dismissive German phrase, *Ostjuden*, Eastern Jews, who spoke Yiddish, worked with their hands, had little secular education, and were far more orthodox in observance. The Russian Jews formed a proletariat in American cities which went on to play a significant role in radical politics and the trade union movement.

By 1917 the campaign against immigrants was in full-flood. An older anti-Catholicism fueled the attack on immigrants from Italy and Sicily. The American Protective Association demanded stricter naturalization requirements. Henry Ford's *Dearborn Independent* blamed Wall Street and the Jews for the difficulties experienced by the American economy. The Ku Klux Klan joined in the attack upon ignorant and illiterate foreigners flooding into the country. The American Federation of Labor demanded literacy tests to restrict the immigration of unskilled labour. A Harvard-educated poet in 1919 labelled a singularly unattractive personage in a poem, "Viennese Chicago Semite", as one might mark a creature in a zoo.

Politicians responded to this outcry with an investigation of the immigration problem, an imposing 42-volume report issued under the chairmanship of Senator William Dillingham. The fact that in 1910 about 72 per cent of the foreign-born lived in urban communities, compared to about 46 per cent for the nation as a whole, was taken to prove that the "new" immigrants were unassimilable. The "experts" who provided the intellectual framework of the Dillingham report were purveyors of racism disguised as "science". The picture they presented of the immigrants was rooted in observable truths (that, in preferring to settle in distinct areas in cities, they were forming ethnic ghettos), but failed to place the decisions made by immigrants as to where they settled either into an economic context or a cultural one. The farm population of the United States was shrinking. The 1890 census revealed that the frontier had "closed". Good new land was no longer available for settlement. New jobs were being created in the factories in the northern and midwestern cities. The cultural factors behind the immigrant ghettos were no deep mystery, nor were they based upon a foreign perversity: the immigrants largely settled where they had friends, relatives or fellow-countrymen, and where their native tongue was spoken. Ethnic identity in the Polish and Bohemian districts of Chicago was supported by the commu-

Immigrants were among the targets of the prohibition laws of the 1920s. Attempts to restrict immigration, and the Volstead Act which enforced the ban on the manufacture or sale of alcohol, seemed to express the same white Anglo-Saxon Protestant hostility towards immigrants and their customs. Immigrants played a significant part in the evasion of prohibition, either through the "cooking" of industrial alcohol in bathroom stills, or through their membership of the organized gangs which smuggled whiskey across the border with Canada, or from "Rum Row" off the Long Island shore. Despite widely-publicized seizures of smuggled alcohol (right: *New York State Troopers unload captured booze in 1920), the government never put sufficient resources into enforcement to do more than modestly raise the price charged to speakeasy customers.*

The two faces of racism. On the left blacks flee a violent race riot in Atlanta, Georgia. On the right, in a saloon in San Francisco, California, clubs and fists are raised against the presence of Japanese.

nal institutions, ethnic newspapers, churches and social rituals. Viewing immigrants as dangerous to American values, the Dillingham report failed to see how true to the American dream—the Horatio Alger dream—of hard work and self-improvement the immigrants were.

The Dillingham report pointed to the difficulty of assimilating the newer groups, and advocated restrictions upon further immigration. The Johnson-Reed Immigration Act, passed in 1924, cut the number of new immigrants to no more than two per cent of the total foreign-born population of that nationality which had been present in the United States in 1890. The "national origins" quota system, established by the Johnson-Reed bill, remained in force until the passage of the Hart-Celler Act in 1965. Immigration restriction reduced the proportion of foreign-born residents in the United States by half. The number of Italian immigrants fell from 222,000 in 1921 to 8,000 in 1926. The intention of such quotas was explicitly racial. Nowhere was the impact of Johnson-Reed felt more strongly than in New York, which had traditionally been the most "foreign" of all American cities. In 1910 over 40 per cent of the city's population was foreign-born. The 1950 census revealed that just under 24 per cent of New Yorkers were foreign-born. (That figure had risen to 28 per cent in the 1990 census.)

The immigrants did not long remain passive dwellers within urban ghettos. Indeed, the ethnic composition of city districts was subject to sharp changes. The Germans who lived in the lower East Side in New York in the 1860s had largely been displaced by Russian and Polish-Jewish immigrants by 1900. They in turn have been replaced by the Hispanic immigrants who currently make the lower East Side their home. We need to think of these ethnic enclaves, not as fixed places in urban life, but temporary staging posts in the histories of families and communities. Increasing prosperity, social mobility and also changes in the job market, lay behind the changing face of an urban people. What changes more slowly and more painfully is the bigotry and racism which greeted the immigrants, tried to restrict their numbers and which feared their presence would destroy "American" values. Some, though perhaps not all, immigrant communities have escaped the worst effects of this bias. But the bias remains and the black community confronts it still.

Urban Life

The new immigration after 1880 and the growth of cities transformed urban life in America.

"Prosperity never before imagined, power never yet wielded by man, speed never reached by anything but a meteor ..."
Henry Adams on New York City, 1904

Between 1880 and 1890 the population of Boston increased from 362,839 to 448,477. In this period over 150,000 familes moved to the city, and nearly 139,000 households left it. There was an astonishing population fluidity in one of the most stable and "mature" American urban areas. About a third of Boston's population was born outside the United States; as young males were heavily over-represented in the immigrant stream, the foreign-born represented 45 per cent or more of the adult male workforce. By the turn of the century, urban America had a new working class, one which was predominantly foreign, and spoke English with difficulty (or not at all).

There were sharper differences between the rural and urban populations. In the largest cities, 70 per cent of white heads of households had foreign-born parents. The figure for Greater New York City was 84 per cent. For rural America, the figure was barely 35 per cent, and was only 42 per cent for the nation as a whole.

The population of New York City had grown forty-fold in the 19th century. But the growth was due almost entirely to immigration from abroad. The Irish and Germans before the Civil War, (and then after 1880), the vast migrations from southern Italy, Russia and Eastern Europe, filled the tenements, provided hands for the gangs which built the elevated railways, dug the subways, built the Brooklyn Bridge and kept the sewing machines humming in the sweatshops on the lower East Side.

War and immigration restriction slowed the flow of new arrivals. The density of population in the tenement districts of New York, which had been among the highest in the world, began to ease. Better housing was to be found along the subway lines to the outer districts of the city, and then the suburbs. By the 1920s, escape from the inner city, and the dream of upward social mobility, became a reality for many urban dwellers.

Emigration may have been a step in the life of people who were already mobile. Leaving the land, the search for industrial employment and the occasional return to the land, may have been repeated once the emigrant reached the United States. But the symbolism of the voyage and the feeling of strangeness on arrival in New York (Castle Garden in the background) were never forgotten.

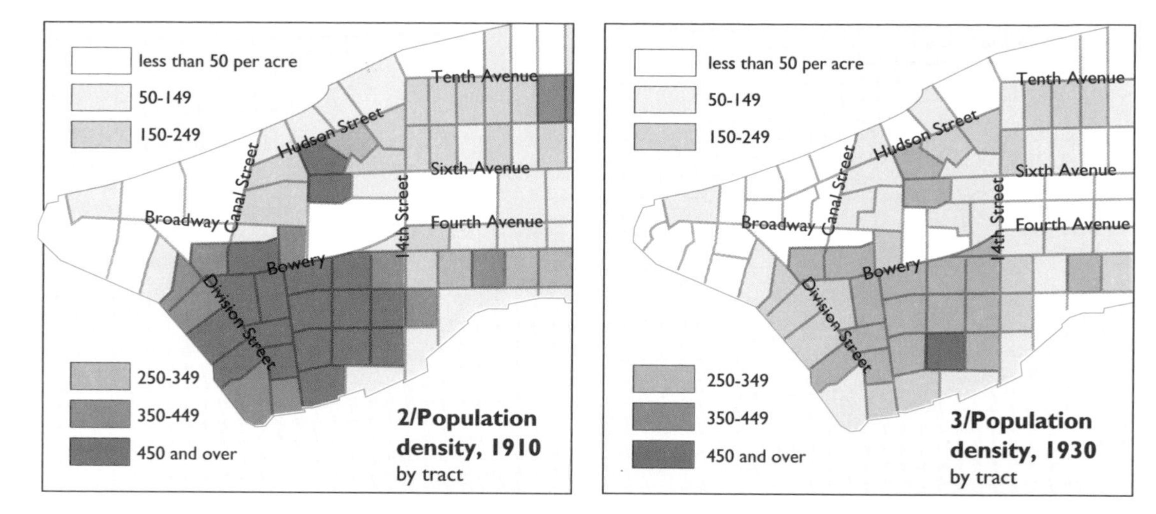

An abandoned harbour fort, Castle Garden served as a public hall until it became the chief US immigrant station in 1855. Over the next 35 years, until it was replaced by Ellis Island, Castle Garden received 7,690,606 mmigrants.

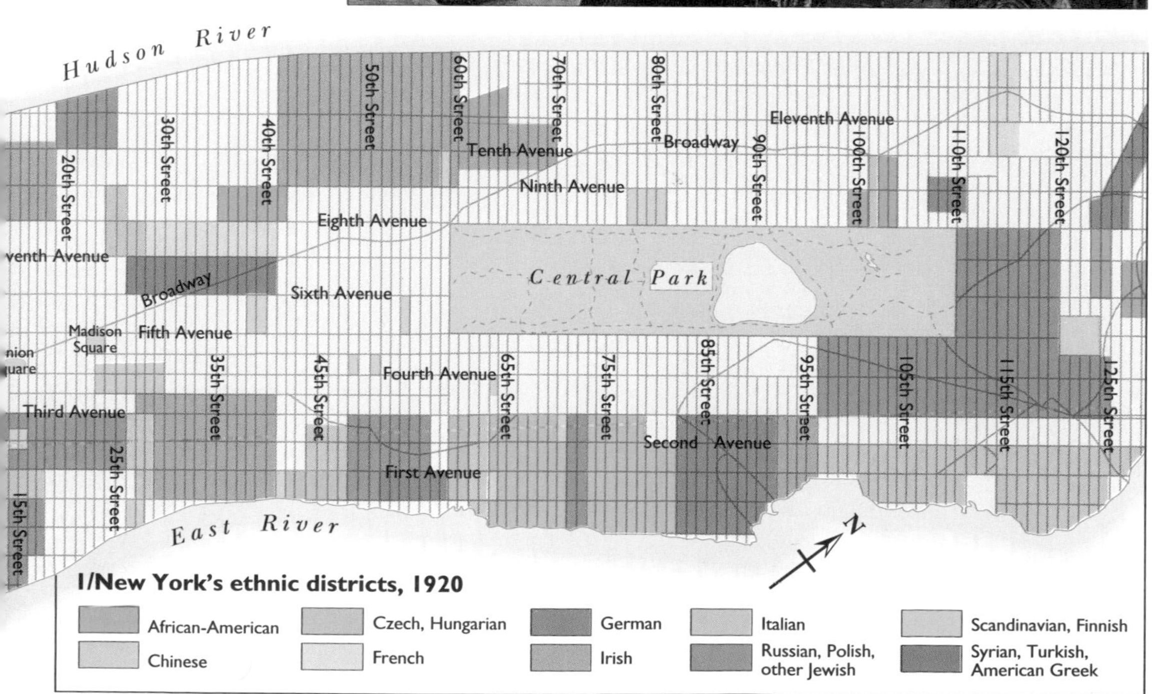

Mexican Revolution

A rebellion against the dictator Díaz in 1910 ended with the democratic Mexican constitution of 1917.

"They will take up their guns with reckless bravery and resist us, desperately in the streets and at the doors of their houses."
John Reed on the Mexican response to the occupation of Veracruz, 1914

Between 1904 and 1913, Mexican oil production rose from over 220,000 barrels to nearly 26 million. British and American firms unambiguously dominated strategic sectors of the economy, including smelting and mining, sugar, rubber and railways. The Mexican dictator, Porfirio Díaz, shaped policy on landowning, labour and foreign investment to suit the interests of foreign capital.

The revolution in Mexico began in 1910 with the overthrow of Díaz by Francisco Madero. By February 1913, with the murder of Madero, American support swung reluctantly behind General Victoriano Huerta as the most likely to restore constitutional government.

In March 1913, supporters of Madero raised the banner of rebellion in the name of constitutionalist government in northern Mexico, led by the governor of Coahuila State, Venustiano Carranza. Huerta's tyranny and threat of civil war persuaded President Wilson of the need to intervene and restore constitutional government. Emiliano Zapata in Morelos in central southern Mexico, and Francisco Pancho Villa in Chihuahua in the north, were leading a revolt by land-hungry *peons* whose support Carranza needed.

In October 1913, after the fall of Torréon to the Constitutionalists, America abandoned Huerta. The harsh arrest of Americans at Tampico gave the Wilson administration a pretext for intervention. On 21 April 1914, 1000 U S Marines landed unopposed at Veracruz. But the vociferous anti-American response across

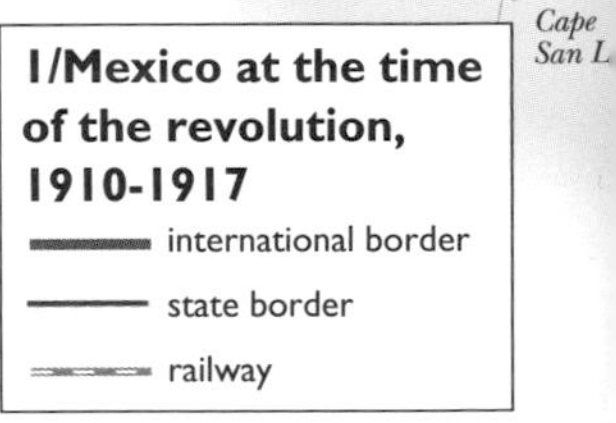

Below: *The nightmare photograph of Mexican conservatives: the leader of the peasant revolt in the south, Zapata, holding a sombrero, confers with the leader of the uprising in the north, Pancho Villa, in general's uniform. Taken in the Presidential Palace, Mexico City, 1914.*

Above right: *The Mexican revolution was constantly agitated by peasant rebellions. This scene, captured by an artist for a French periodical, appeared in 1924.*

Mexico shocked the Americans. In July the pressure on Huerta succeeded, and he resigned. Mexico City was taken by the Constitutionalists.

The Revolution scarcely paused before devouring its own; Carranza's forces attacked Pancho Villa in April and May 1915. Villa's men were reduced in 1916 to bandit raids across the United States' border. U S Army cavalry, under the command of General John Pershing, chased Villa across northern Mexico. The Revolution settled down. Carranza promulgated a democratic constitution in February 1917. Two months later, America entered the war in Europe.

The First World War

"There should really be no neutrals in a conflict like this ..."
American serving in the French Foreign Legion, October 1915

The "war to end war" drew America into the heart of the stubborn conflicts of Europe.

The unexpected news in August 1914 of war in Europe left the American people unshaken. They were neutrals, proudly aloof from the struggles of the old world. Public opinion was outraged by stories of war atrocities committed in Belgium by the invading Germans; an influential group of politicians felt that a victorious Germany would pose a grave threat to American interests. But it was not "our" war.

The task of British propaganda was to persuade the Americans otherwise. Appeals to a common cultural heritage, and a humanitarian concern for the sufferings of wartime Europe, swayed opinion towards the Allies.

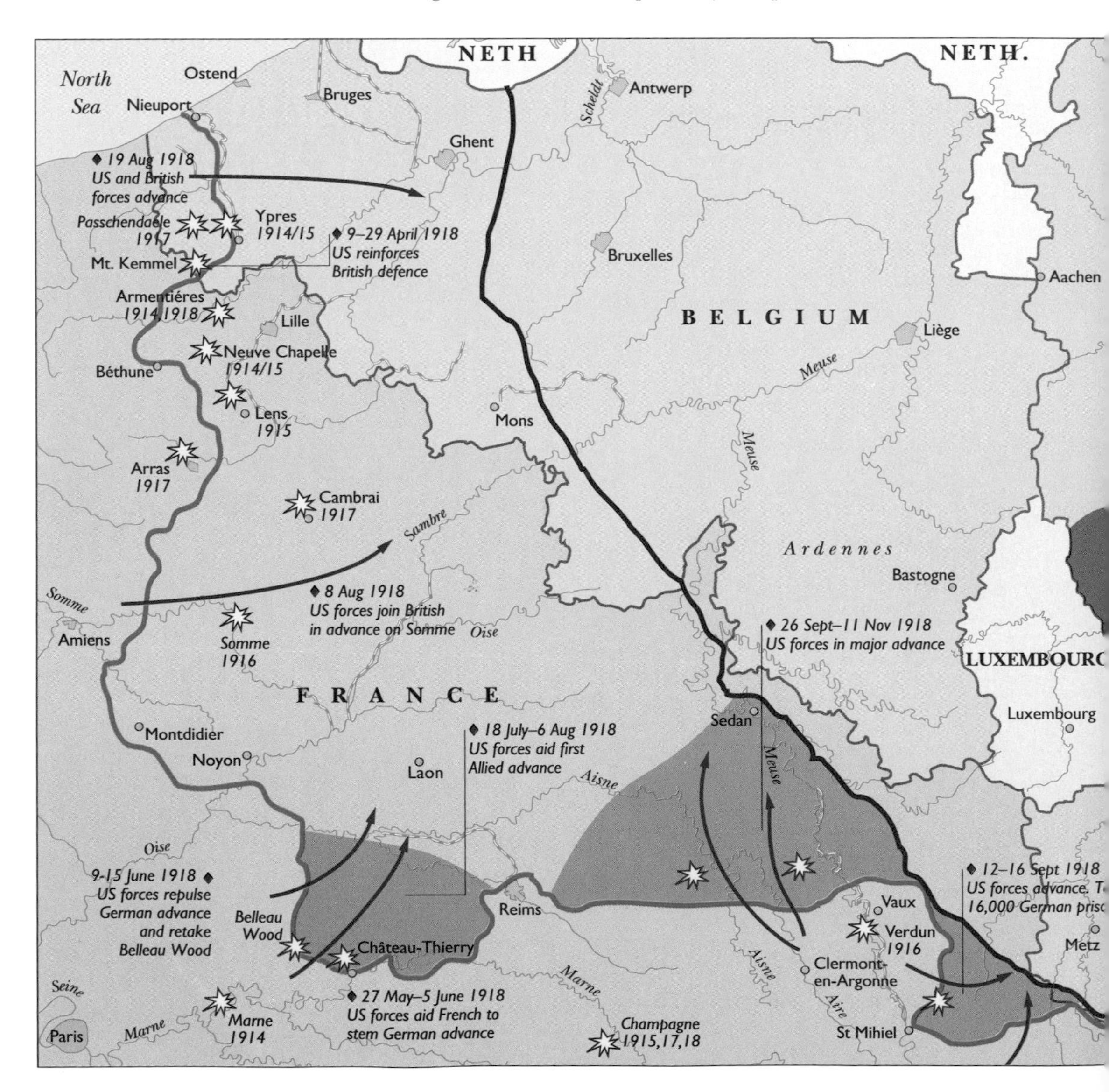

Far right: *President Wilson (1856–1924) was the last American president to have been born before the Civil War. He was a powerful, reforming president, who articulated American ideals. It was the compromises of the peace, not the war, which undid him.*

Right: *Tin-pan Alley led the wave of patriotic sentiment which followed President Wilson's declaration of war on 2 April 1917. Vaudeville shows featured flag-waving songs and sharp mockery of the German Kaiser.*

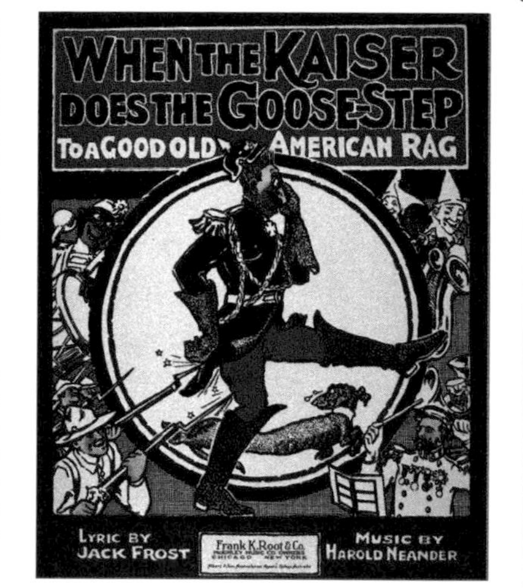

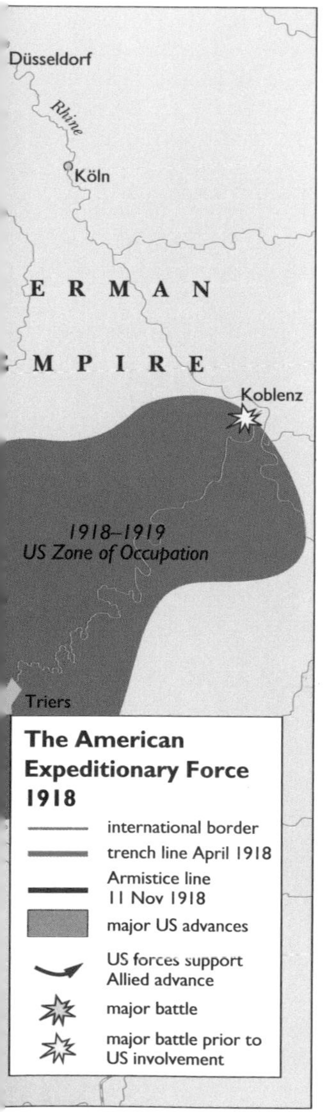

The arrival of the United States 2nd Division north of the Aisne on 3 June 1918 blocked a German advance along the Chemin des Dames. On 6 June 1918 the 2nd Division launched the first major American action in the war, with 27,500 soldiers and Marines recapturing Balleau Wood, Vaux and Bouresches.

"Preparedness" campaigns sought to bring America into the war on the side of Britain and France. The sinking of the British liner Lusitania on 7 May, 1915, with a loss of 785 passengers (including 128 Americans) and 413 crew, stunned the nation. Public opinion indignantly demanded recognition of the rights of neutrals.

President Woodrow Wilson was re-elected in 1916 as the man who kept the United States out of the war. But early in 1917 the German decision to recommence unrestricted submarine warfare made neutrality impossible. America declared war on 6 April. By the war's end in November, 1918, some 2 million American soldiers had been sent to Europe, of which 50,000 had died. The cost of the war was $33 billion. A War Industries Board was created to regulate the national transportation system and organize and control the economy. The experiment in a controlled and planned economy was abandoned once the war was over, but the success of government propaganda was not forgotten—nor was the experience of economic planning.

The war for America was brief, victorious and produced a handsome bevy of war heroes. Real wages rose by 50 per cent from 1914 to 1918, and membership of the patriotic American Federation of Labor rose sharply.

The Making of Harlem

Migration from the rural South to the industrial cities of the North and Midwest made Black Americans the most heavily urban of American ethnic groups.

"The pulse of the Negro has begun to beat in Harlem." Alain Locke, *The New Negro*, 1925

In 1990 nine out of ten blacks lived below the Mason-Dixon line, the unofficial border between North and South. Fewer than 20 per cent of black southerners lived in cities. "Jim Crow" laws (the legal basis for racial segregation in the South), the boll weevil—which devastated the cotton fields after 1901— and the disenfranchisement of black voters by grossly discriminatory electoral regulations, shut the door on hopes that conditions for blacks were going to improve in the South.

There had been small, stable black populations in most northern cities. Despite the bigotry and condescension of whites, opportunities for education and employment in northern cities was better than in the South, although there was very little upward social mobility. Nonetheless, from the 1890s there was a sustained migration of blacks from the South into northern cities. They travelled the Illinois Central Railroad from Mississippi to Chicago, and the steamships from Norfolk, Virginia, to New York. By 1910 the black population of New York had risen to 91,709; in 30 years, it reached 458,000.

A rim of "border cities" were often initial migration targets and staging posts for the further move to larger cities like Chicago, Philadelphia and New York. In each of these cities a small black population was dramatically increased and in each a "black ghetto" of effectively segregated housing and schooling emerged. Northern cities were not the "Promised Land", but con-

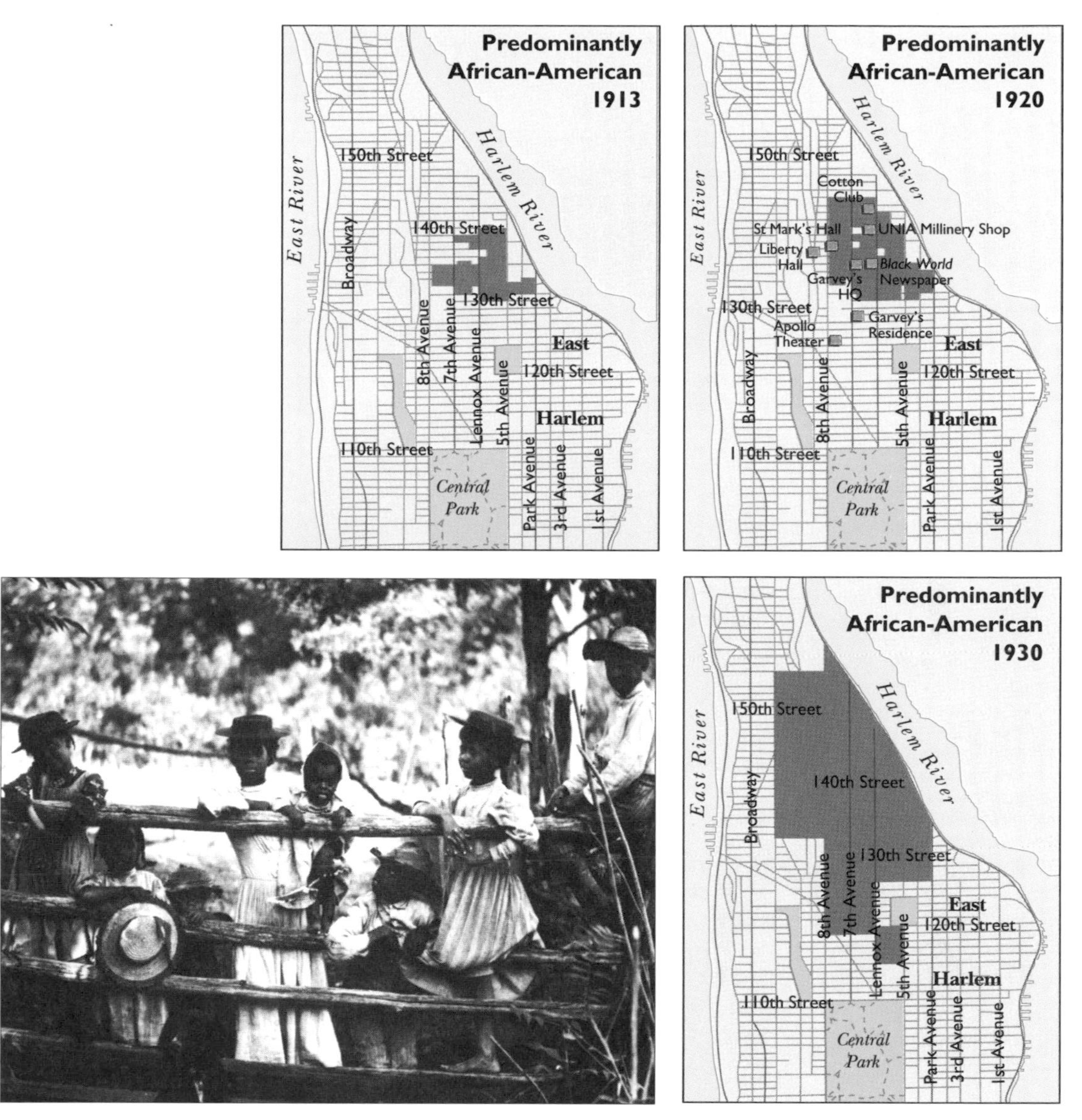

The difficult decision to leave the South and seek a better life in the big cities in the North and Midwest changed the black experience in the United States. Migration was nothing new: northward migration had been going on since the Civil War. Blacks from the eastern coastal states, from Virginia to Florida, headed for the east coast cities. Alabama blacks headed for Ohio and Michigan. Blacks from the Mississippi Valley followed the Illinois Central line from New Orleans to Chicago. The Second World War opened a new stream as blacks from eastern Texas and Louisiana headed for California.

ditions were better than in the rural south, and the migration continued.

The first black families moved into Harlem in 1905. Harlem, long a semi-rural retreat for aristocratic New Yorkers, was certainly no slum. The post-1918 housing boom created more vacancies in Harlem, which soon became a magnet for blacks. By the 1920s, it was the largest black community in the nation and home to 70 per cent of the city's black residents.

The urban experience shaped the "New Negro" who emerged in Harlem.

> "I've been a worker:
> Under my hand the pyramids arose.
> I made mortar for the Woolworth Building."
> (Langston Hughes, "Negro", 1926).

In Harlem there was a remarkable self-discovery out of which the Harlem Renaissance was born. Jazz, parties and show business—as well as alienation, anger and rage—were ingredients in the urban pressure cooker from which new struggles and new self-definitions were born.

The Industrial Colossus

The emergence of large corporations and trusts brought to an end the era of individual enterprise in the American economy.

"The business of America is business."
President Calvin Coolidge, 1925

The movement towards combinations, "arrangements", and "trusts" rapidly extended in the 1870s throughout the economy. Whether through "horizontal" integration (the acquiring of direct competitors), or "vertical" integration (either "backward" through the acquisition or control of the supply of raw materials, or "forward", seeking control of transportation and marketing), the future lay in the large combinations. What Rockefeller's Standard Oil Company achieved in oil refining, Andrew Carnegie sought in steel, Pillsbury in flour milling. The results were impressive: the United States surpassed Britain in steel output in the 1880s. J.P. Morgan purchased Carnegie Steel in 1901 for $480 million. It was the largest sale in American industrial history to that date. Carnegie remarked that "the day of individual competition in large affairs is past and gone".

Frederick Winslow Taylor preached the gospel of "scientific management": breaking complex tasks down to their simplest components. The skilled worker who took 290 minutes to assemble a fly-wheel magneto at a Ford plant was replaced by an assembly line of men who took just 13 minutes for each item. The life of an assembly-line worker became more stressful, but the manufacturer benefited from increased efficiency and lower costs.

When the United States entered the war in 1917, large companies could produce the needed weapons, ships and war matériel. In the 1920s, the favourable climate for industry continued. Profits rose; membership of trades unions declined. But by 1929 the runaway advance of stock prices lost contact with the real value of corporate assets. "Wall Street Lays an Egg" was the *Variety* Magazine headline after the October 1929 crash.

Top right: *Gerrit A. Beneker's 'The Builder' was a capitalist version of the images in the Soviet Union heroically portraying the proletarian worker, and paintings of square-jawed Aryan model-workers in Nazi Germany. The actual workers in America, with their disconcerting enthusiasm for trades union membership and industrial militancy, were seldom treated with as much dignity as in Beneker's image.*

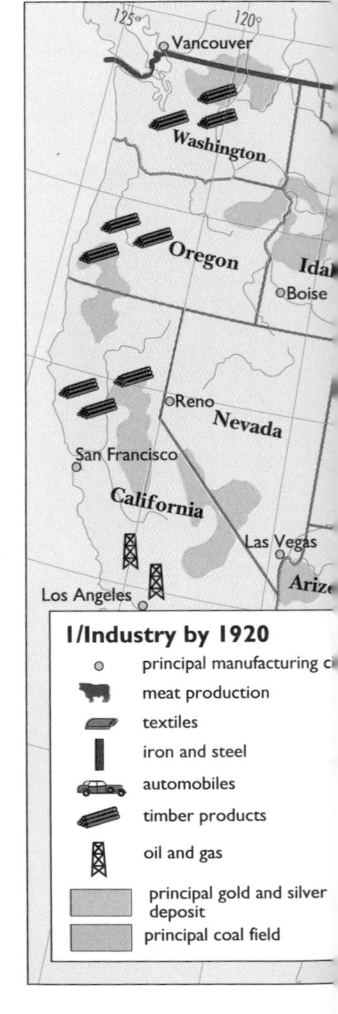

Below: *Thomas Hart Benton's mural "America Today" was painted in 1930–31 for the New School for Social Research, New York City. The panels, which represent many common scenes from American life, were 7 feet 6 inches high and between 8 feet and 26 feet wide.*

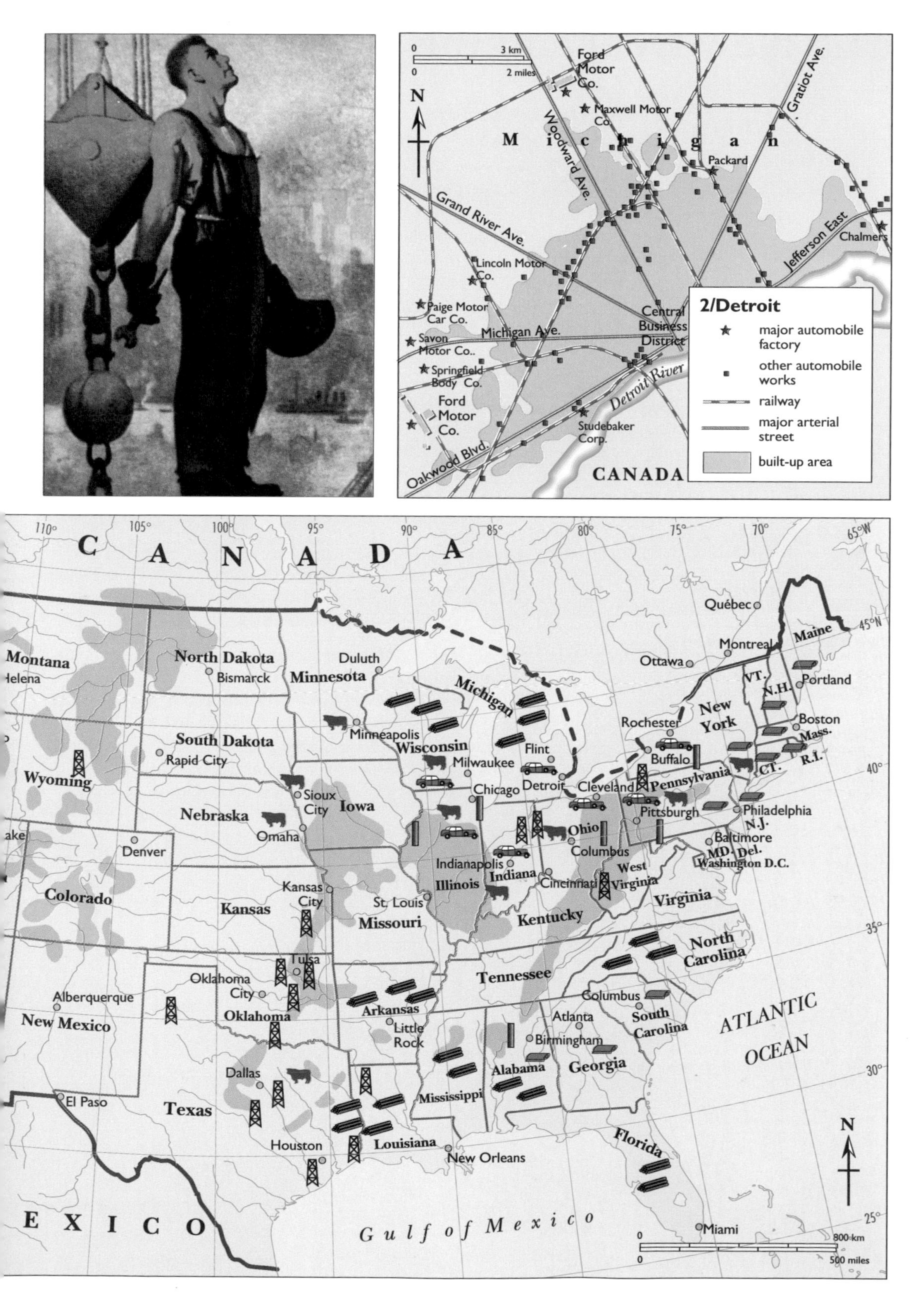
2/Detroit
major automobile factory
other automobile works
railway
major arterial street
built-up area
0 3 km
0 2 miles
N
Michigan
Ford Motor Co.
Maxwell Motor Co.
Woodward Ave.
Gratiot Ave.
Packard
Grand River Ave.
Jefferson East
Chalmers
Lincoln Motor Co.
Paige Motor Car Co.
Savon Motor Co..
Michigan Ave.
Central Business District
Springfield Body Co.
Ford Motor Co.
Detroit River
Studebaker Corp.
Oakwood Blvd.
CANADA
110°
105°
100°
95°
90°
85°
80°
75°
70°
65°W
45°N
40°
35°
30°
25°
CANADA
Québec
Montreal
Ottawa
Maine
Montana
Helena
North Dakota
Bismarck
Duluth
Minnesota
Michigan
VT.
N.H.
Portland
New York
Rochester
Boston
Mass.
R.I.
CT.
South Dakota
Rapid City
Minneapolis
Wisconsin
Milwaukee
Flint
Buffalo
Wyoming
Sioux City
Iowa
Chicago
Detroit
Cleveland
Pennsylvania
Pittsburgh
Philadelphia
N.J.
Nebraska
Omaha
Ohio
Columbus
Baltimore
MD.
Del.
Washington D.C.
Denver
Colorado
Kansas
Kansas City
Indianapolis
Indiana
Illinois
Cincinnati
West Virginia
Virginia
St. Louis
Missouri
Kentucky
North Carolina
Tulsa
Tennessee
Oklahoma City
Oklahoma
Arkansas
Little Rock
Columbus
South Carolina
Albuquerque
New Mexico
Atlanta
Birmingham
Georgia
Alabama
ATLANTIC OCEAN
Dallas
Texas
El Paso
Mississippi
Houston
Louisiana
New Orleans
Florida
N
MEXICO
Gulf of Mexico
Miami
0 800 km
0 500 miles

The New Deal and the TVA

"This is pre-eminently the time to speak the truth ..."
Franklin Delano Roosevelt in his inaugural address, 3 March 1933

Top right: ***"Marching for the NRA", a song inspired by the National Recovery Administration system of 1933 which was the New Deal agency at the front line of the struggle to halt the spiraling deflation of the Depression.***

Below: ***the famous Norris Dam in East Tennessee was constructed by the TVA. The projects of the CCC (Civilian Conservation Corps), with as many as 500,000 young men working on reforestation, road construction, national park and flood prevention programmes, were important symbols of the New Deal and its approach to the problem of unemployment.***

President Roosevelt and the New Deal offered hope to Americans at the depths of the 1930s depression.

No matter how it was measured—in terms of stock prices, rising car sales or the quantity of booze available in the local speakeasies—the 1920s was a decade of sustained self-satisfaction and patriotism. The attack on trades unions was presented by industry as the "American Plan". Advertisers had a new product to sell: the idea of consumerism itself. "You do not sell a man the tea," remarked a senior copywriter in 1925, "but the magical spell which is brewed nowhere else but in a teapot". The 1920s was a "magical spell" which came to a hard end.

What made the Depression so demoralizing was that from 1929 to 1933 every effort at economic stabilization resulted in a lower plateau. The per cent unemployment figure of 1929 (3.2 per cent) rose remorselessly during President Herbert Hoover's years in office. By 1933 the official figure had reached 24.9 per cent. Total steel output fell from 100 in 1929 to 41 in 1933. Cash receipts from farming were halved. Farm wages kept in step with the decline.

The defeat of Hoover in the 1932 presidential election by Franklin Roosevelt, Democratic Governor of New York, brought to the White House a wily, effective politician who was blessed with a remarkable gift for communicating with the public. He gave Americans hope.

The New Deal, a broad mix of interventionist programmes designed to use the federal government—not to change or reconstruct American society—but to preserve the democratic order, save capitalism and provide relief from hardship. The Tennessee Valley Authority, established in 1933, was given the brief to develop agriculture and industry in the Tennessee Valley, to produce electricity and improve flood control. Shying away from the word "plan", the TVA nonetheless was a bold symbol of new social energy and an impressive demonstration of what a society might achieve if only it had the political will.

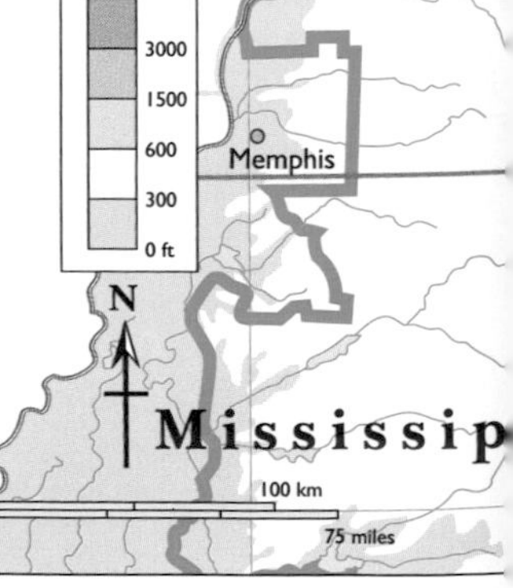

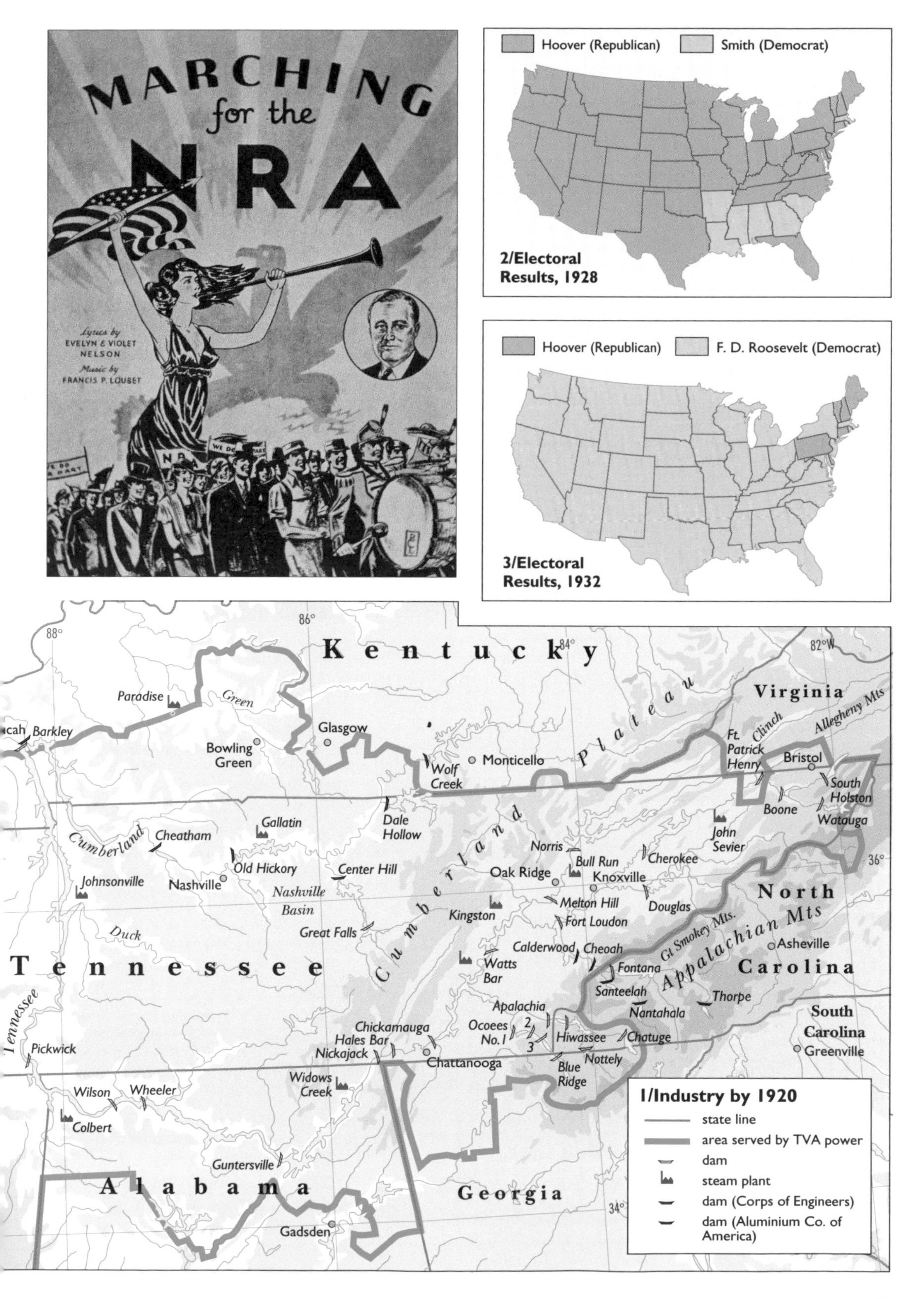
MARCHING for the NRA
Lyrics by EVELYN & VIOLET NELSON
Music by FRANCIS P. LOUBET
Hoover (Republican)
Smith (Democrat)
2/Electoral Results, 1928
Hoover (Republican)
F. D. Roosevelt (Democrat)
3/Electoral Results, 1932
Kentucky
Virginia
North Carolina
South Carolina
Tennessee
Alabama
Georgia
Cumberland Plateau
Appalachian Mts
Gt Smokey Mts.
Allegheny Mts
Nashville Basin
Paradise
Barkley
Bowling Green
Glasgow
Monticello
Wolf Creek
Dale Hollow
Gallatin
Cheatham
Old Hickory
Center Hill
Johnsonville
Nashville
Great Falls
Kingston
Oak Ridge
Norris
Bull Run
Knoxville
Cherokee
Melton Hill
Fort Loudon
Douglas
John Sevier
Ft. Patrick Henry
Bristol
Boone
South Holston
Watauga
Calderwood
Cheoah
Watts Bar
Fontana
Santeelah
Nantahala
Thorpe
Asheville
Apalachia
Ocoees No. 1
Hiwassee
Chatuge
Nottely
Blue Ridge
Chickamauga
Hales Bar
Nickajack
Chattanooga
Widows Creek
Pickwick
Wilson
Wheeler
Colbert
Guntersville
Gadsden
Greenville
1/Industry by 1920
state line
area served by TVA power
dam
steam plant
dam (Corps of Engineers)
dam (Aluminium Co. of America)

Tenements

In 1900 one and a half million New Yorkers lived in tenements. Housing conditions in the city were the worst in the world.

"Pneumonia, Typhoid, and Influenza ran up and down the icy tenement halls, playing deadly tricks like schoolboys on a lark."
Michael Gold, *Jews Without Money*, 1930

The pressures which created the skyscraper (increased land values and ever-expanding demand for space in a crowded urban environment) lay behind the tenements which spread like dank mould across American cities. With each newly-arrived immigrant vessel from Hamburg or Liverpool the population grew remorselessly. More housing was needed.

Older homes in less-fashionable parts of the city, usually three or four storeys high, began in the 1840s to be divided into accommodation for two families on each floor. In the back garden, a small colonial-era stable or workshop, might be rebuilt to house tenants. The profits to be made, even from the poorest immigrants, supported some of the city's wealthiest families. The Vestry of Trinity Church on fashionable Broadway, and John Jacob Astor, owned large parcels of land in the slums. Agents handled the rentals; it was in their interests to "pack" as many tenants into the properties as possible. Conditions deteriorated in the tenements which became flea-ridden, dirty and a hazard to public health. As the home of prostitution and illicit gambling, they were also regarded as a moral threat to the welfare of the city.

Aware that the tenement was a spectacularly good investment, developers began to erect larger and larger structures, built with cheaper materials and to ever-lower standards. (Gotham Court on Cherry Street in New York, perhaps the worst tenement in the United States, housed 504 persons.) Apartments had no heating. Running water was only available in the filthy hall. Toilets were located in the basement or backyard, and served by inadequate drains. The odour of the larger tenements was unforgettable.

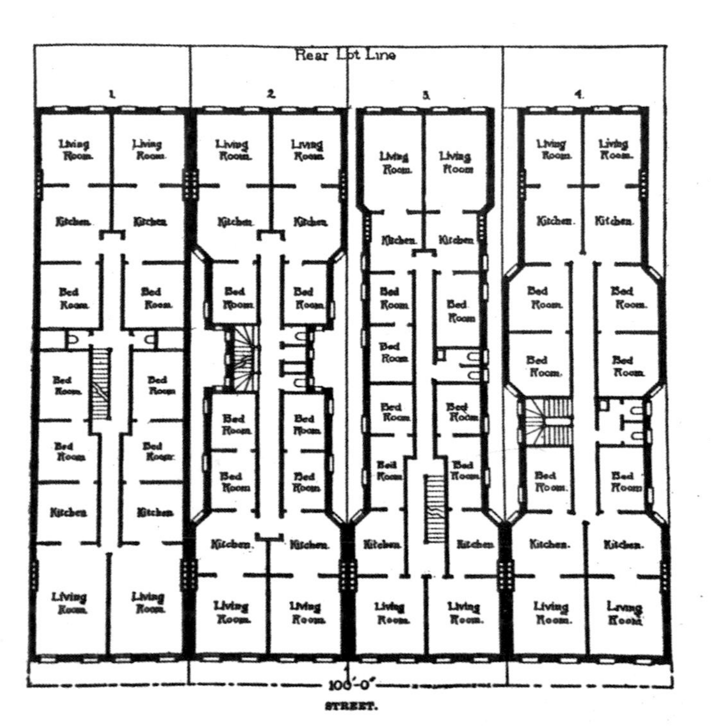

Philanthropists sought to construct model tenements, upon enlightened design principles. The provision of water closets on each floor was one such improvement, but in most tenement apartments the "dark" sleeping rooms remained without access to daylight or ventilation. Damp mattresses and stale air encouraged the diseases that were rampant in urban tenements.

At heart, the problem lay in the chaotic *laissez-faire* nature of the housing market. Reformers in New York and other major urban centres sought to regulate conditions in tenements by tightening the building code. The "Old Law" tenement (permitted by legislation passed in 1879), created the "dumbbell" with its sunless vertical airshaft. "New Law" tenements (built

Reformers hoped that improving the design of tenements would solve the social and sanitary problems of slum housing. The narrow air shaft (floor plan, far left *and cut-away view of a tenement,* left) *theoretically brought air and light to interior "dark" rooms. But conditions continued to deteriorate.*

after 1901, which had better ventilation and separate toilets), were decided improvements. The older structures remained.

Tenements came to symbolize the worst aspects of the dismal legacy of free-market urban housing. Regulation achieved some valuable reforms, but it was not until the New Deal that the Federal Housing Agency encouraged the wholesale clearing of inner-city slums, and the demolition of the tenements and their replacement by well-designed housing projects funded by the city, state and federal governments. The many errors of that policy will remain with urban America well into the next century.

VI: The Imperial Mantle

Wartime planning and co-operation shaped the post-war world—until mutual trust collapsed, and the Cold War began.

"The greatest nation on earth either justifies or surrenders its leadership. We must choose ..."
Arthur H. Vandenberg, Republican Senator from Michigan

The "American Century"

At the end of the war in 1945 the American people looked to the future with confidence. Success in war was based upon the fabled American industrial and agricultural strength, and upon the wartime anti-fascist alliance with the Soviet Union. Americans believed that the post-war world would be constructed out of New Deal programmes (Social Security Act, passed August 1935), and a strong commitment to internationalism. The new international order took shape during the war. A stable system of exchange rates and the International Monetary Fund were set up at the Bretton Woods Conference in New Hampshire in July 1944. The Dumbarton Oaks Plan (Washington, D.C., August–October 1944) served as the basis for the Charter of the United Nations (San Francisco, April–June 1945).

Wartime structures also played a crucial role in post-war reconstruction. The U N Relief and Rehabilitation Administration (established in November 1943) was the vehicle through which the United States provided Europe with over $11 billion in aid and loans in the immediate aftermath of

A C-47 (Dakota, landing left) is dwarfed by the C-74 "Globemaster" being unloaded on the runway of Gatow Airport, Berlin, on 19 August 1948. Capable of carrying 20 tons of flour, the "Globemaster" carried automobiles belonging to American military personnel on the return flight to Frankfurt. The U.S. Air Force used Dakotas and Globemasters to overcome the blockade which the Russians imposed on Berlin in June 1948.

the war. The Marshall Plan (proposed by Secretary of State, General George C. Marshall, at an address at Harvard University in June 1947) got off to a rocky start when Molotov, the Soviet Foreign Minister, walked out of the preliminary meeting, charging that the plan was an "imperialist" plot. In December 1947 President Truman submitted a request to Congress for $17 billion to fund the European Recovery Program. After a heated debate, the Marshall Plan package won support in Congress in April 1948, and an initial funding of $5.3 billion. By 1950, the total American funding reached $12 billion.

The meaning of American generosity changed sharply in the aftermath of the war. The Potsdam Conference, held in July 1945, brought the two new western leaders, the British Prime Minister Clement Attlee, and Present Truman, into negotiations with Stalin. Potsdam maintained the outward form of mutual trust. Agreement was achieved about occupation zones in Germany, de-Nazification and the establishment of provisional borders between Germany and Poland. But the mutual confidence required to sustain these policies was shaken, and then destroyed by unilateral actions which reflected neither the basic principles of consultation or the cooperation which sustained the wartime alliance.

Cold War

Stalin, who was a cynical realist, believed he knew the deeper intention of the capitalist states bordering the Soviet Union. There was to be no soften-

Enjoying their commanding officer's sense of humour are (left to right) *Vice-Admiral Arthur Struble, Brig. Gen. E.K. Wright, and Maj. Gen. Edward Almond. In the chair: General Douglas MacArthur (1880–1964), commander-in-chief of the United Nations Forces in Korea. This photograph was taken in Inchon harbour, in the aftermath of the Allied landing on 15 September 1950. The Marines and Republic of Korea forces had retaken Seoul, the capital of South Korea by the end of the month.*

ing of the war he conducted against the Communist Party or the peoples of the Soviet Union. The purges of the 1930s were resumed at the end of the war, sweeping tens of thousands of soldiers into the Siberian Gulags. A new wave of repression, and violent demands for ideological orthodoxy, were launched against Russian writers, artists and musicians. Cooperation over Germany ceased. Communist parties in Eastern Europe, riding the legacy of anti-fascism and wartime resistance, were on the verge of power. Free elections were blocked or subverted. Both Greece and Turkey were put under pressure by the Soviet Union. By 1948 Czechoslovakia was the last of the Western European states to install Communists in government. Were the Russians intending to dominate all the Eastern European territories which had been liberated by the Red Army? Interpretations differed. Among the most influential was the analysis of George Kennan, attaché in the U S Embassy in Moscow. In February 1946 he sent a "long telegram" to the Secretary of State:

> "In summary, we have here [in the Soviet Union] a political force committed fanatically to the belief that with the US there can be no permanent *modus vivendi,* that it is desirable and necessary that the internal harmony of our society be disrupted,

Many of the leading Hollywood personalities protested the HCUA hearings in 1947. A group of 500 left Hollywood to travel to Washington to protest the hearings. From left to right (front row)*: Geraldine Brooks, June Havoc, Marsha Hunt, Lauren Bacall, Richard Conte and Evelyn Keyes.* Back row*: Paul Henreid, Humphrey Bogart (spokesman), Gene Kelly and Danny Kaye.*

> our traditional way of life be destroyed, the international authority of our state be broken."

Kennan called for the Soviet state to be studied with "courage, detachment [and] objectivity", and proposed that the public be educated in the realities of the Soviet situation. Of course American internal problems would have to be addressed, but what above all was needed was an American policy guided by a "positive and constructive" engagement with post-war reconstruction. A return to isolationism was inconceivable. A month later Winston Churchill, speaking at Fulton College in Missouri, gave the post-war world its abiding metaphor: "From Stettin in the Baltic to Trieste in the Adriatic, an iron curtain has descended across the continent."

The European powers were in no position to resist Communist pressure in Greece. It fell to President Truman to accept the challenge. Early in 1946, his intention to stiffen the American response to the Soviet Union was expressed to aides: "Unless Russia is faced with an iron fist and strong language, another war is in the making. Only one language do they understand—'how many divisions have you?' I do not think we should play at compromise any longer." The "Truman Doctrine", set out in March 1947, enshrined the defence of the free world and the "containment" of Russian power as the objective of American policy. By 1947 economic and military aid to Europe no longer only expressed American generosity and idealism: it was a calculated assessment of American interests.

The Marshall Plan did not immediately sweep away the long-standing American caution at "entangling alliances". But the wartime model of planning and cooperation created the framework within which the economic reconstruction of Europe became part of the broader confrontation with Communism. The Berlin Airlift, from June 1948 to May 1949, amounting to 300,000 flights carrying 2.5 million tons of supplies, was a practical expression of the determination of the British and American governments to maintain the Allied position in Berlin. At the height of the Berlin blockade, the new Secretary of State, Dean Acheson, responded to the request of European states by forming the North Atlantic Treaty Organisation. The mutual-defence enshrined in NATO was followed by a pattern of commitments and pacts which spanned the globe. The economic supremacy of the United States, and a broad consensus which blamed isolationism for the outbreak of the war in 1939, made the far-flung commitments of post-war policy generally acceptable to the American people.

Fidel Castro (1927–) was the leader of the revolution which successfully overthrew the corrupt Cuban dictatorship of Fulgencio Batista. For Americans, Castro was the most dangerous of all the Third World Communist leaders, and repeated attempts were made to assassinate Castro and topple his regime.

The War Within

The Cold War was a domestic battle no less than an international one. The American war against Communism could be dated back to 1918, and the 1941–1945 wartime alliance with the Soviet Union was perhaps only a hiatus in a longer enmity. The "Palmer Raids" and deportations of alien radicals 1919–1920 were launched by Attorney General A. Mitchell Palmer and conducted by the Immigration Bureau, Military Intelligence and the General Intelligence Division (headed by J. Edgar Hoover) of the Department of Justice. Arrests and swift deportations (later declared unconstitutional) obliterated the Anarchist movement. Raids and prosecutions at the same time destroyed the Industrial Workers of the World (the "Wobblies"). The combined effects of World War I and the Bolshevik revolution in Russia split and then maimed the Socialist Party in the United States. By 1919, it was little more than a shadow of the mass socialist parties of Europe. In the public mind the Communist Party of the United States (founded in 1919 by radicals breaking away from the Socialist Party) was perceived as being extremist, disloyal and guilty of subversion. The destruction of the Communist leadership in trades unions in the 1930s did much to tame the union movement. Social welfare and public ownership, which formed the

The Rev. Martin Luther King (1929–1968) was a Baptist minister and advocate of non-violent action for civil rights for American blacks. He played a major role in the March on Washington in 1963, which brought 200,000 demonstrators to the nation's capital. He is seen here with Ralph Abernethy (left) *on a demonstration with the United Nations flag in the background. When he began to question the war in Vietnam, the FBI began to take a serious covert interest in Dr. King..*

social democratic agenda in Europe, was regarded as being un-American. A new, non-socialist language had to be found in the Depression to describe such welfare measures.

The wartime alliance with the Soviet Union was wildly popular, and sympathetic journalistic portrayals of the Red Army, Soviet commanding officers and Communist economic might were commonplace. The film industry enthusiastically portrayed the Russian contribution to the war effort. The key films, all of which appeared in 1943–1945, are *Mission to Moscow*, directed by Michael Curtiz, based upon the memoirs of the United States' Ambassador to the Soviet Union in the late 1930s; *North Star*, a Lewis Milestone film with a script by Lillian Hellman; and *Song of Russia* by Gregory Ratoff about pianists, guerrilla fighters and so on. Of these, *Mission to Moscow* was praised as "a magnificent propaganda exercise". What came to haunt Hollywood was the film's attempt to defend Stalin, and to justify the 1937 purge trials from the Stalinist point of view.

Anti-Communism moved from the margins of public consciousness to the centre after the war. In 1945 Congressman Martin Dies' Special Committee on Un-American Activities was given permanent status, and a new title, as a

The postwar American highway programme was funded as a defence measure. The resulting interstate system linked the highway programmes of individual states, led by California and New York. The most populous states, with the highest numbers of cars registered, both confronted significant problems relating the transportation needs of highly built-up urban areas with the great distances which separated the urban areas. The elegance of American road and bridge design was for many decades the envy of the world. The separation of express from local traffic, and the soaring flyovers entering and leaving the freeway system in Los Angeles (above) *transform what are little more than utilitarian structures into a brilliant symphony of movement and structure.*

standing committee. J. Edgar Hoover, Director of the Federal Bureau of Investigation since the 1920s, warned of Communist secret espionage in America. This, and the pressure of events abroad, combined to link the external and internal threats. Only 10 days in March 1947 separated the promulgation of the Truman Doctrine and the issuing of an executive order by President Truman which established a loyalty and security programme for all federal employees, and at the same time revived the Attorney General's list of subversive organizations.

"Unfriendly" witnesses

The entertainment industry was a "soft" target, with a promise of abundant publicity for ambitious right-wing Congressmen. The investigators found the heads of Hollywood studios very helpful. The first phase of hearings into "un-American" activities in Hollywood began in September 1947 when 41 witnesses were subpoenaed. The background of these events was a period of labour unrest and strikes in the Hollywood studios, and an increasing tendency to use the label of Communism to discredit studio unions. Among the early witnesses was Adolphe Menjou, who told the Committee that Hollywood had become one of the main centres of red activity, and that numerous films had secretly been influenced by Communists in the industry. Out of the first group of witnesses, a small group of 19 planned to take a stand against the Committee. Unwilling to "name names" (of others in the industry who were Communists or "fellow travellers"), they were regarded as "unfriendly". At the hearings in October 1947, 10 of those subpoenaed refused to answer the Committee's questions. They were cited by the House of Representatives for contempt in November, and afterwards indicted by a Grand Jury. In April 1948, the playwright John Howard Lawson and the novelist Dalton Trumbo were convicted of contempt and sent to prison.

The studio bosses met at the Waldorf-Astoria Hotel in New York on November 24, 1947, and formulated a collective policy to save the studios at the expense of those of their employees who were in trouble with the House Committee (HCUA). The studio bosses agreed to cooperate fully with HCUA, and to blacklist those like the Hollywood Ten who had refused to testify. Dalton Trumbo and the writer Lester Cole were immediately sacked by MGM; Ring Lardner was fired by 20th-Century Fox; the director Edward Dmytryk was fired by RKO. Dore Schary, the Head of Production at RKO, who had a reputation as a leftist, informed the Screen Writer's Guild that in order to protect the freedom of the industry, it would henceforth be necessary to fire and blacklist the Ten and to avoid hiring anyone believed to be a Communist. The studios led the way in destroying the careers of the unfriendly Ten. John Houston, Humphrey Bogart, Gene Kelly and William Wyler formed a Committee for the First Amendment, and arranged for two programmes defending the Ten to be broadcast by ABC. Those who supported the Ten soon found themselves the object of blacklisting by the studios. Of 204 people who signed a friend-of-the-court brief to the Supreme Court on behalf of the Ten, no fewer than 84 were blacklisted.

The second phase of HCUA investigations took place in 1951, when 45 "unfriendlies" were subpoenaed. The witnesses' tactic had altered in response to the conviction of the Ten. They now decided to decline to answer particular questions on the ground that the Fifth Amendment guaranteed individuals the right not to incriminate themselves by their own testimony. By taking the Fifth, they escaped prosecution for contempt, but nonetheless faced the destruction of their careers. The Screen Writer's Guild, which had defended the Hollywood Ten, withdrew support for the blacklisted writers in 1953, and a year later introduced a loyalty oath and barred Communists from membership.

The "Red Scare" or anti-Communism, from its inception in the politics of wartime America in 1917 until the ending of the Hollywood blacklists in the 1960s, was a crucial subtext for American culture. To a large extent, the anti-Communism made good political sense for its exponents. Careers of conservatives like those of Richard Nixon, Joseph McCarthy and Ronald Reagan flourished in its heyday. In that environment, Liberals found that combative anti-Communism was virtually their only hope of political survival. Anti-Communism was the shaping belief of 20th-century American politics.

Anti-Communism lay at the heart of the commitments which brought the United States to Korea, the Bay of Pigs and to Vietnam. The real cost of the Cold War is perhaps incalculable. During the Cold War the real "true believers" fervently looked for signs of the collapse of the Soviet system. Realists sought détente and co-existence. Obscure nationalists talked about ethnic rivalries which would resurface when Russian control over Eastern Europe ended. In the West, political élites admired the "Prague Spring" of Alexander Dubček, and hoped in time that a socialism with a "human face" and a general lowering of the ideological temperature would make life more bearable for what were no longer (in polite circles) referred to as "the captive peoples" of Eastern Europe. Sophisticated politicians in the West doubted whether the unification of Germany was possible. Some argued, for domestic consumption, that a divided Germany was better for the rest of Europe. Everyone was taken by surprise when the collapse came.

The imperial mantle hangs awkwardly from the shoulders of a weakened, not overly confident United States. There has been little genuine celebration of a staggering historical triumph. The winners of this great ideological duel look at the cost of victory, and at the state of America in the 1990s, and wonder whether one more such "triumph" will finish off the nation for good.

Marines take cover after landing from a helicopter in Quang Ngai Province in August 1967. The helicopter transformed the conduct of military operations, adding an ability to resupply and move troops which had seldom been available to World War II armies. The helicopter used for medical evacuation, and advances in trauma medicine, also helped to keep casualty levels to an historic minimum. Nonetheless, over 50,000 Americans were killed in Vietnam.

World War II: The Pacific

The devastating Japanese attack on Pearl Harbor brought the United States into the war.

"Here lie three Americans ..." Caption for a photograph by George Strock, *Life* Magazine, 20 September 1943

When war broke out in Europe in September 1939, President Roosevelt urged Congress to repeal the Neutrality Acts of 1935–37, which banned arms sales to belligerent powers. Large new expenditures for national defence in 1940, and sales of arms, ammunition and ships to Great Britain (expanded in 1941 under the Lend-Lease Act) made it clear that the United States would not remain disengaged. Retaliation for the banning of exports of scrap iron and steel to Japan, and the freezing of Japanese bank credits after the occupation of French Indo-China, led to the Japanese air attack on the naval base at Pearl Harbor in Hawaii on Sunday, 7 December, 1941. War was declared on the 8th.

Warfare in the Pacific involved naval battles in which surface ships never made contact with each other, and fought with carrier-based planes. The

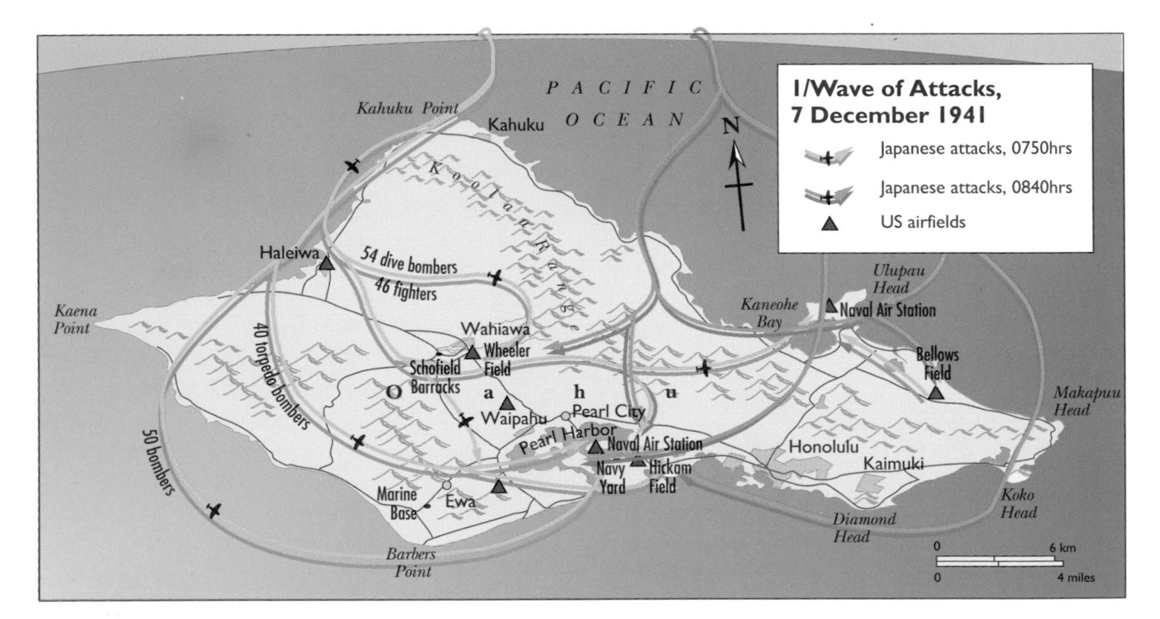

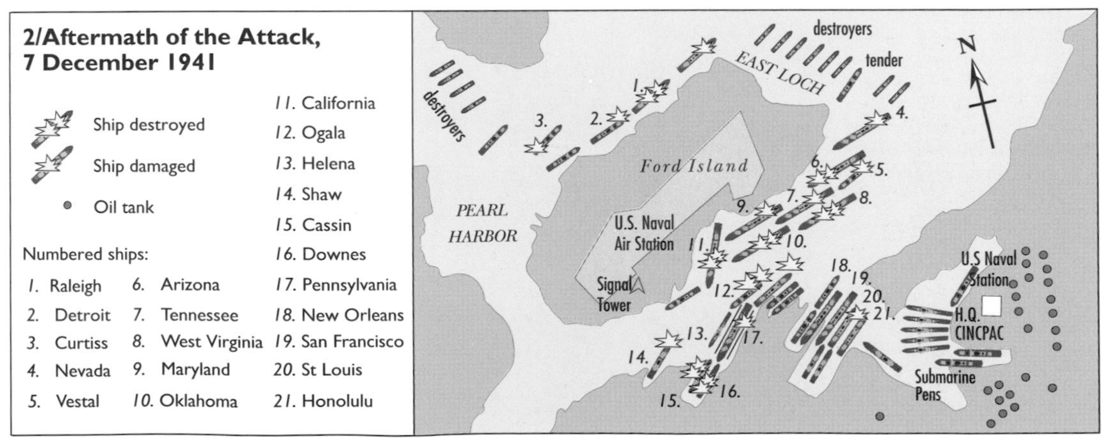

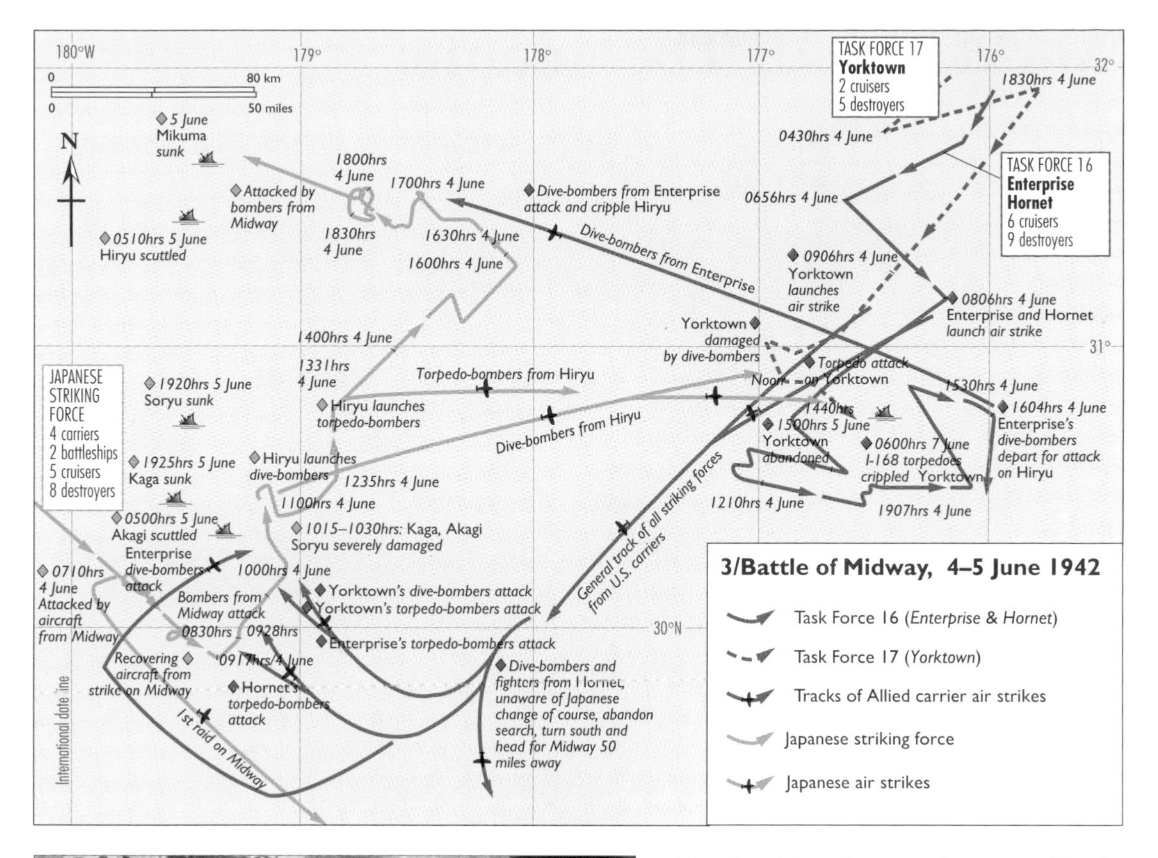

island and jungle campaigns were fought with a unique savagery, and with an edge of explicit racial hatred which could only be compared to the war on the Eastern Front between Germany and Russia.

It was not until the battle of Midway in June, 1942, that the seemingly inexorable advance of Japanese arms was halted. Island-hopping invasions, thunderous naval engagements, and deadly submarine warfare brought the Allies to Okinawa in March, 1945. Only 360 miles southwest of Japan, the American air force sought to end the war without invasion. The fire-bombing of Tokyo by 334 bombers under General Curtis LeMay on 8 March, 1945, left nearly 84,000 dead and a fourth of the city's buildings gutted. The war was ended with the dropping of an atomic bomb on Hiroshima on 6 August, and on Nagasaki on 9th August. There were 41,322 American dead out of 170,596 casualties in the Pacific war.

The newsreels and photographs of the smashed American fleet in Pearl Harbor aroused public opinion to fever-pitch and took the United States into the war.

World War II: The Atlantic

Priority was given to war in Europe, and to the defeat of Nazi Germany.

"I hate war."
President Franklin D. Roosevelt, 14 August 1936

The decision of Germany and Italy on 11 December, 1941, to declare war on the United States ended the "European" war, and began the global or world war which ended in 1945 with Rome abandoned by the retreating Germans, and Berlin in ruins—occupied and sacked by the Red Army.

The aggressive militarism of Japan deeply troubled the Americans, but they shared the British belief that Hitler posed the greatest world-wide threat. In secret UK–USA staff talks in 1941, a Europe-first policy for the war was agreed.

The lightning German victories of 1939 could not be sustained. The failure of the *Luftwaffe* in the Battle of Britain, the first thousand bomber raid on Cologne (30 May, 1943), Montgomery's victory at El Alamein (29 June, 1943), and then the stunning Russian counter-offensive at Stalingrad (19 November, 1943), left the war stalemated.

American soldiers under Major General George S. Patton joined the British in the encircling of Rommel's Afrika Korps in Tunisia. The invasion of Sicily, then Italy, followed—campaigns which proved bloodier and more difficult than anyone expected. But the opening of a "Second Front" to relieve the German pressure on the Russians could not be achieved until Operation *Overlord* began on 6 June, 1944. Four thousand landing craft carried 176,000 troops, the largest amphibious operation in history, to the coast of Normandy. Paris was liberated by 25 August, but the defeat of the British at Arnhem, and the German counter-attack in the Ardennes (Battle of the Bulge, 16–26 December, 1944) delayed the ending of the war by six months.

Out of 12,466,000 Americans enrolled in the armed forces, 322,188 were killed (all theatres), and 700,000 wounded.

The Normandy landing was above all a triumph of Allied industrial and logistical organization. Good weather on D-Day, 6 June 1944, also helped. G.I.s wade ashore, a small part of the 176,000 troops landed by 4000 invasion craft. Operation Overlord *was the largest amphibious operation in history. By 2 July the Allies had placed ashore at Normandy about one million troops, 171,532 vehicles and landed 566,648 tons of supplies.*

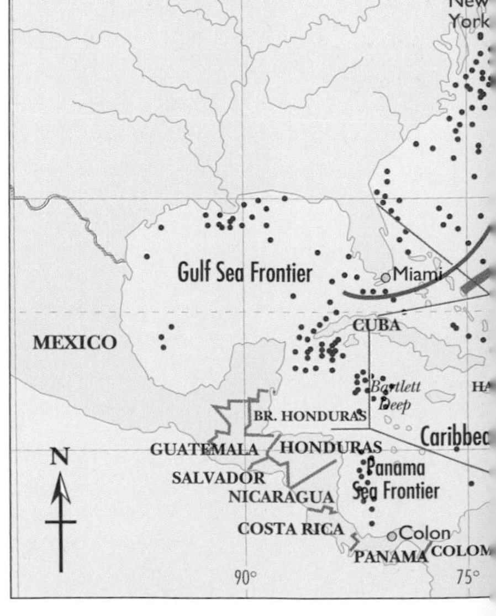

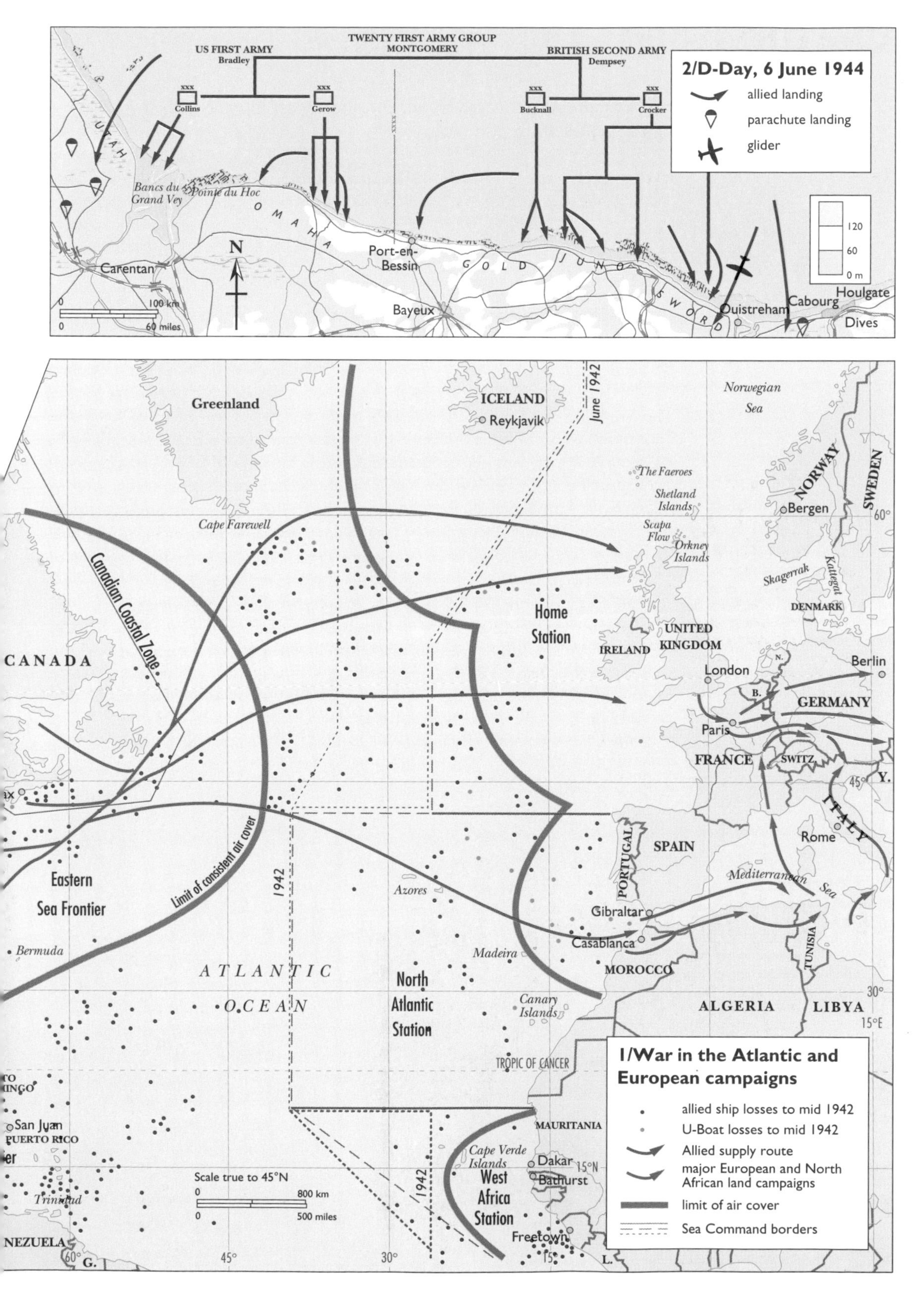
2/D-Day, 6 June 1944
allied landing
parachute landing
glider
TWENTY FIRST ARMY GROUP
MONTGOMERY
US FIRST ARMY
Bradley
BRITISH SECOND ARMY
Dempsey
Collins
Gerow
Bucknall
Crocker
UTAH
OMAHA
GOLD
JUNO
SWORD
Bancs du Grand Vey
Pointe du Hoc
Carentan
Port-en-Bessin
Bayeux
Ouistreham
Cabourg
Houlgate
Dives
1/War in the Atlantic and European campaigns
allied ship losses to mid 1942
U-Boat losses to mid 1942
Allied supply route
major European and North African land campaigns
limit of air cover
Sea Command borders
Greenland
ICELAND
Reykjavik
Norwegian Sea
The Faeroes
Shetland Islands
Scapa Flow
Orkney Islands
NORWAY
SWEDEN
Bergen
Skagerrak
Kattegat
DENMARK
Cape Farewell
Canadian Coastal Zone
Home Station
IRELAND
UNITED KINGDOM
London
Berlin
GERMANY
Paris
FRANCE
SWITZ.
ITALY
Rome
SPAIN
PORTUGAL
Mediterranean Sea
Gibraltar
Casablanca
MOROCCO
ALGERIA
LIBYA
TUNISIA
CANADA
Eastern Sea Frontier
Limit of consistent air cover
Bermuda
Azores
Madeira
Canary Islands
ATLANTIC OCEAN
North Atlantic Station
TROPIC OF CANCER
MAURITANIA
Cape Verde Islands
Dakar
Bathurst
West Africa Station
Freetown
San Juan
PUERTO RICO
Trinidad
Scale true to 45°N
800 km
500 miles
June 1942
1942

On the Road

When the automobile came, the American people moved faster and further than ever before.

"... a family's motor indicated its social rank as precisely as the grades of the peerage determined the rank of an English family."
Sinclair Lewis, *Babbitt*, 1922

Long before the invention of the automobile, American culture was rich with poets who praised personal freedom and mobility:

Afoot and light-hearted, I take to the open road,
Healthy, free, the world before me,
The long brown path before me, leading wherever I choose.

Walt Whitman, "Song of the Open Road", *Leaves of Grass* (1856).

Politicians preached about the virtues of what Theodore Roosevelt called "the strenuous life". Advice to "Go West, young man!" (Horace Greeley) found a ready audience. Across the continent Americans exhibited a restless hunger to move on. There was an American gift for improvisation, for the spontaneously created community of wagon-trains crossing the Great Plains. The slow, steady business of building a civic order held fewer attractions. It was a society of people on the move, ready to drop everything at the rumour of gold in California, or the chance of cheap land in the West. Farmsteads were abandoned; mining towns boomed in the West only to become ghost towns a few years later. When the Dust Bowl ruined agriculture in Oklahoma, Ma and Pa Joad in John Steinbeck's *Grapes of Wrath* (1939) loaded the kids in their old jalopy and headed for California. The stage coach, train and then automobile in turn became enshrined in the mythology of a culture. Millions of man-hours were spent lovingly discussing differentials and the new Chevy. Automobile sales rose from 2,300,000 in 1922 to 4,500,000 in 1929. Mobility tended to break down the provincialism of society, or at least spread provincialism across the continent. The automobile brought an expectation of better roads, (US1—map right—was an early example of improved highway standards) and higher standards of service. It also led to motels, Dunkin' Donuts and Howard Johnson restaurants.

The automobile shown in this photograph "A sunny day at Milford, Connecticut", (taken by T.S. Bronson in 1908), was among the first examples of 20th century technology to be widely used by women. Women's clothing adjusted to the requirements of travel, with the smart cloche replacing the wide-brimmed bonnet and scarf by the 1920s. Driving gloves, and flat-heeled shoes or boots—which were expected for horseback riding—transferred smoothly to the wardrobe of the well-dressed automobilist.

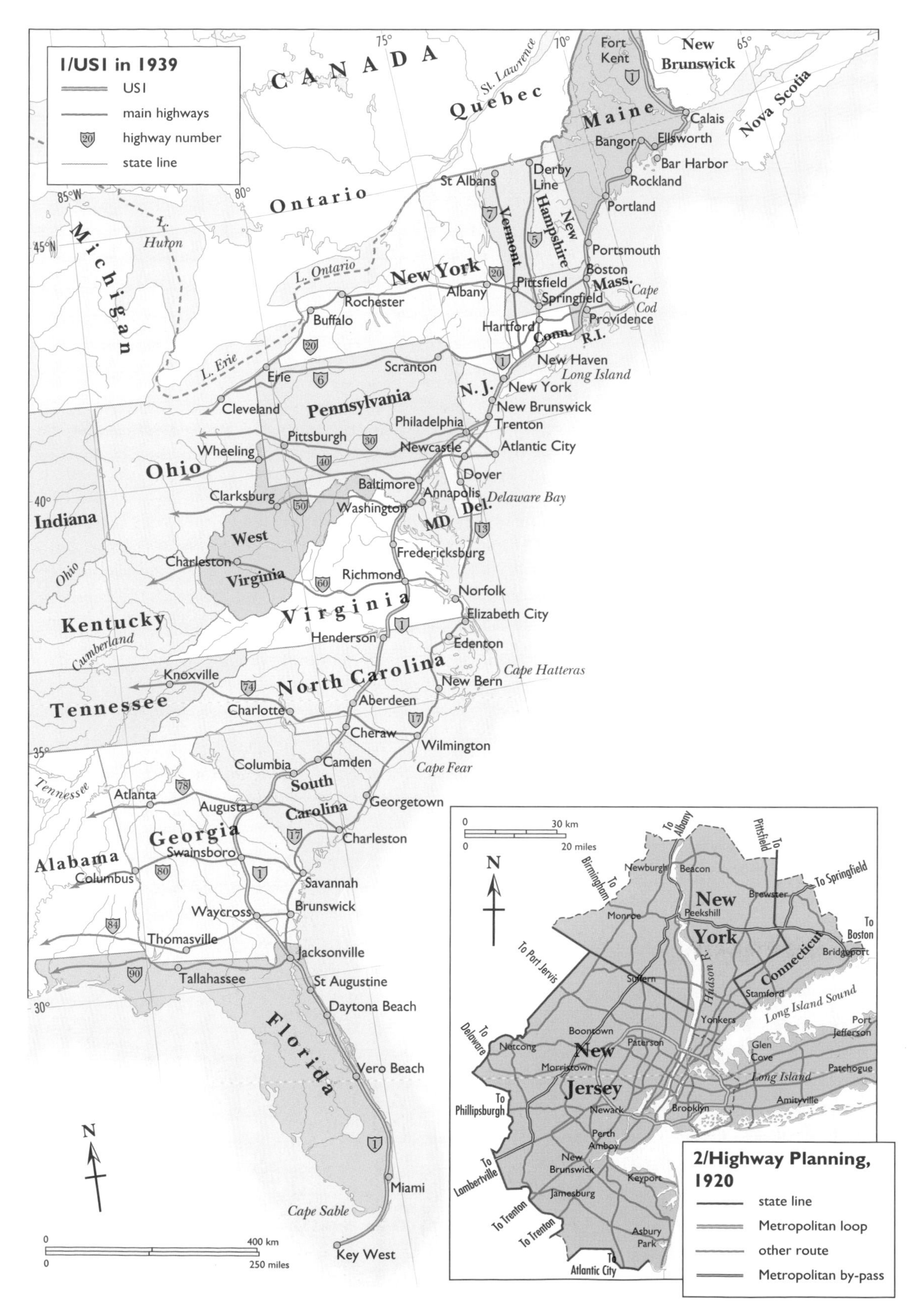
1/USI in 1939
USI
main highways
highway number
state line
CANADA
Quebec
Ontario
Maine
New Brunswick
Nova Scotia
Michigan
New York
Vermont
New Hampshire
Mass.
Conn.
R.I.
N.J.
Pennsylvania
Ohio
Indiana
West Virginia
Virginia
Kentucky
Tennessee
North Carolina
South Carolina
Georgia
Alabama
Florida
MD
Del.
Fort Kent
Calais
Bangor
Ellsworth
Bar Harbor
Rockland
Portland
Portsmouth
Boston
Derby Line
St Albans
Albany
Pittsfield
Springfield
Hartford
Providence
New Haven
Rochester
Buffalo
Erie
Cleveland
Scranton
New York
New Brunswick
Trenton
Philadelphia
Newcastle
Atlantic City
Pittsburgh
Wheeling
Dover
Baltimore
Annapolis
Washington
Clarksburg
Charleston
Fredericksburg
Richmond
Norfolk
Elizabeth City
Henderson
Edenton
New Bern
Knoxville
Charlotte
Aberdeen
Cheraw
Wilmington
Columbia
Camden
Georgetown
Atlanta
Augusta
Charleston
Swainsboro
Columbus
Savannah
Waycross
Brunswick
Thomasville
Jacksonville
Tallahassee
St Augustine
Daytona Beach
Vero Beach
Miami
Key West
Cape Sable
Cape Fear
Cape Hatteras
Delaware Bay
Long Island
Cape Cod
St. Lawrence
L. Huron
L. Ontario
L. Erie
Ohio
Cumberland
Tennessee
75°
70°
65°
80°
85°W
45°N
40°
35°
30°
0
400 km
250 miles
2/Highway Planning, 1920
state line
Metropolitan loop
other route
Metropolitan by-pass
0
30 km
20 miles
New York
New Jersey
Connecticut
Hudson R.
Long Island Sound
Long Island
To Albany
To Pittsfield
To Springfield
To Boston
To Birmingham
To Port Jervis
To Delaware
To Phillipsburgh
To Lambertville
To Trenton
To Trenton
To Atlantic City
Newburgh
Beacon
Brewster
Peekshill
Monroe
Bridgeport
Stamford
Suffern
Yonkers
Port Jefferson
Glen Cove
Patchogue
Amityville
Boontown
Paterson
Netcong
Morristown
Newark
Brooklyn
Perth Amboy
New Brunswick
Keyport
Jamesburg
Asbury Park

Suburb

As the inner core of American cities declined, millions of families sought the good life in suburbia.

"No man who owns his own house and lot can be a Communist. He has too much to do."
William J. Levitt, builder of the Levittown suburban housing developments

The 1980 census revealed that for the first time since 1810, the population of rural America had grown faster than the urban centres. The metropolitan population remained large, and growing, but the rate of growth was faster on the edges, in small towns and suburbs, than in the city centres. This is the shape of post-metropolitan America, with its shopping malls, well-funded schools, and immunity (until the 1980s) from the problems of drugs and violence.

The post-war suburbs were the legacy of a pent-up demand by young families with children for homes at a moderate cost. The Federal Housing Agency supplied large builders with credit and mortgage insurance which enabled prefabricated construction of small single-family suburban dwellings. (The first Levittown was built on Long Island, 30 miles from New York City.) Families on barely 10 per cent above the national median income of $5,000 were able to buy a house with a down payment of $100. Millions of young couples became suburbanites.

Critics of suburbia have concentrated upon the low density of suburban housing, and the lack of diversity or social infrastructure. The suburbs were, until the 1970s, white; they remain thoroughly middle-class. The image of the suburban housewife, proudly smiling before her freshly polished floor as her 2.3 children return from school, is an enduring one, and conveys a cruel truth. Suburbs were machines for the nuclear family; few women with small children worked outside the home. Without nearby parents or relatives, the burdens of parenthood could not be shared. Despite the isolation of suburban life, and the pathology of its sterile tidiness, its place as the American ideal is secure.

Suburban politics have also become proverbial. Having effectively detached themselves from the turmoil and financial problems facing inner cities, suburban dwellers seldom vote for taxes to support crumbling inner-city infrastructure. Politicians, dependent upon suburban votes, behave accordingly.

One of the greatest innovations in modern advertising was the advertiser's discovery that it was not necessary to appeal to the buyer's reason by listing the qualities of a product. Rather, they learned to appeal to the non-rational yearnings of consumers by explaining why—for example—a brand new stove (left) would give Mrs Suburban America a fuller, richer life, a better marriage and a happier family.

Allentown

PENNS

1/The spread of Suburbia into New Jersey by 1980

city–borough
highway
state line
county line
distance from Times Square

The happy family, c.1952. Despite worrying revelations about infidelity in the Kinsey Report, the media image of the American male was a central guiding figure in the family, actively sharing the pleasures of the latest issue of a magazine with handsome, attentive children and a youthful, contented wife.

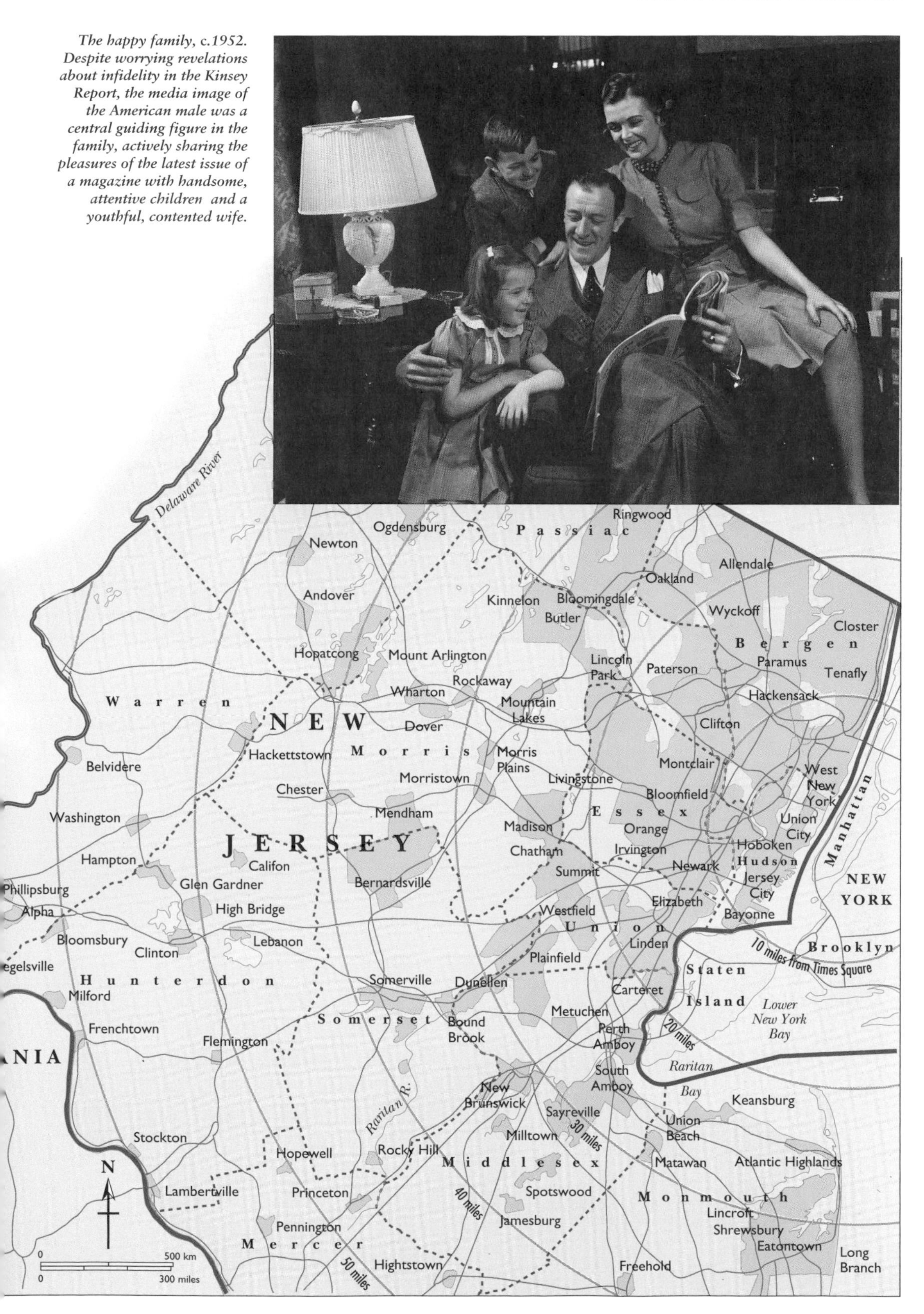

The Cold War

"It became necessary to destroy the town in order to save it."
Unnamed U S Army Major, Ben Tre, Vietnam, February 1968

Conflict between the Soviet Union and the United States and its allies reached across the planet.

The Cold War formally began with the deterioration of relations between the victorious Allies after 1945. Its origins, however, reach back to the American declaration of war in April, 1917. Public opinion supported the decision to go to war, but it was feared that immigrants from Germany and Austria-Hungary, radicals, anarchists and the followers of the Bolsheviks in Russia who had opposed the war, would commit sabotage or weaken public resolve. A "red scare" in the early months of the war silenced opposition periodicals and dissenting opinion generally. Continuing into 1920, the first great "red scare" increasingly focussed upon the Bolsheviks, and their plans for world revolution. All forms of domestic radicalism were tainted with the Red Menace.

A powerful spirit of internationalism and a belief in collective security, supported the wartime alliance with the Soviet Union and the creation of the United Nations in 1945. But it was an alliance which Americans entered into with open eyes. When the mutual suspicions deepened, a "Cold War" extended across the globe.

Angry analogies drawn from appeasement, and the abandonment of Czechoslovakia and Poland in the 1930s, led to the creation of NATO and other regional military alliances, underpinned by the dollar, the H-bomb and the Strategic Air Command. Early warning radar tracked flights of

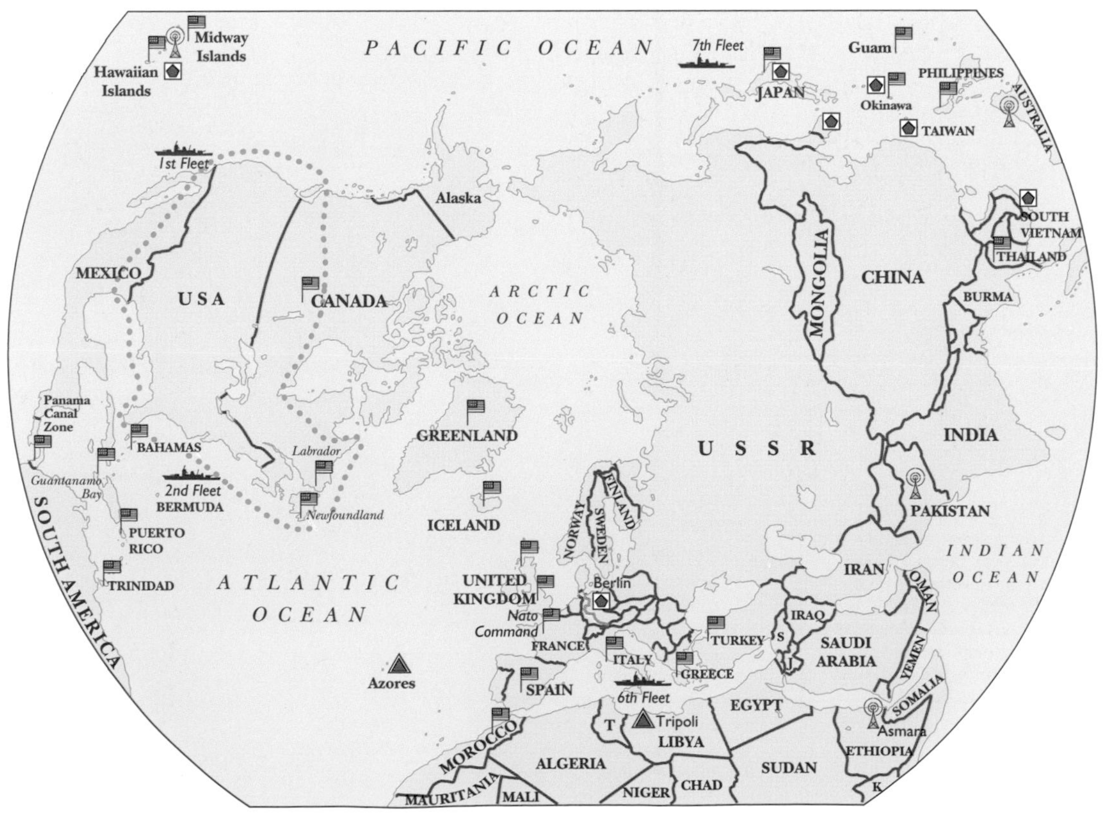

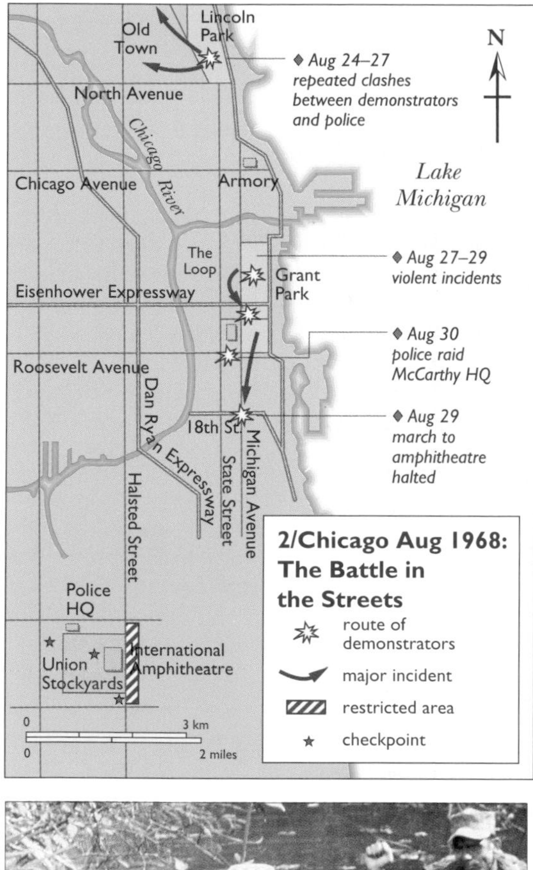

geese in the Arctic. Communist "aggression" was checked in what were wars in everything but name. (Korea was a "police action"; no one quite knew what category Vietnam belonged to.)

There was a domestic price to pay for the vast Cold War arms expenditure. An even more enduring wound was inflicted by the war in Vietnam.. The legacy of a divided society, increasingly doubtful of foreign commitments, meant that there was more relief than cheers and ticker-tape parades when the Cold War was won, and the Soviet Union collapsed in 1989.

The United States created a "Special Forces" unit, known as the Green Berets, specially trained for counter-insurgency and guerrilla conflicts in the Third World. They were strongly supported by President Kennedy, and served with the American forces in Vietnam.

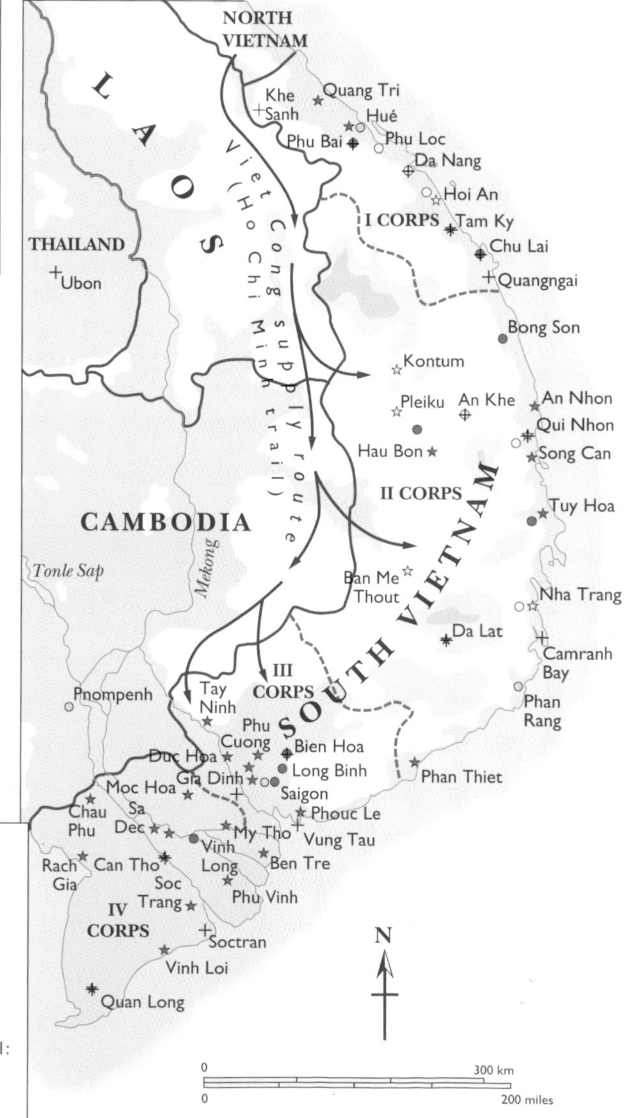

Rich and Poor

The problem of poverty remains the great challenge before American society.

"We're in the money."
Song title by Harry Warren and Al Dubin, 1933

The geography of wealth and of poverty invites specific, local comparisons. The short physical distance between East Harlem and the Upper East Side in Manhattan is perhaps the greatest economic gulf in American society. With a per capita income of nearly $60,000 in the 1990 census, and with 31 per cent of the residents possessing graduate degrees, the Upper East Side ranks with the most privileged areas in the nation. In East Harlem, with 37 per cent of families living below the poverty level, only 1 per cent of residents have higher degrees, and the per capita income in 1990 was $8,888.

The figures shift, decade by decade, but it is no longer clear whether such figures any longer retain the power to stir the concern or moral indignation of the community. The political economist, Henry George, argued in 1885 that "the great majority of those who suffer from poverty are poor not from their own particular fault, but because of conditions imposed by society at large. Therefore, I hold that poverty is a crime—not an individual crime—but a social crime ... for which we all, poor as well as rich, are responsible." The belief that poverty is a great wrong is not often encountered in public discourse today.

Attitudes towards poverty and wealth have changed, sometimes sharply. For most of the last two centuries the acceptance of the American dream had carried with it the possible nightmare outcome of failure, despair and hardship. People have made their choice, and borne up, grumbling but usually stoical, before the consequences. A new deal, a fair deal, a rebalancing of the scale between rich and poor, has regularly been demanded, and sometimes realised. The acceptance of wealth and fortitude before the discomforts of materialism have given to Americans a deep reverence for success and its rewards.

Attempts to boost morale during the Depression produced feel-good advertising campaigns (from 1937). Photographers loved these billboard images, for real life—in this case, victims of a flood in Louisville, Kentucky, lining up for Red Cross relief—repeatedly offered ironic perspectives on the grinning middle-class family in their automobile.

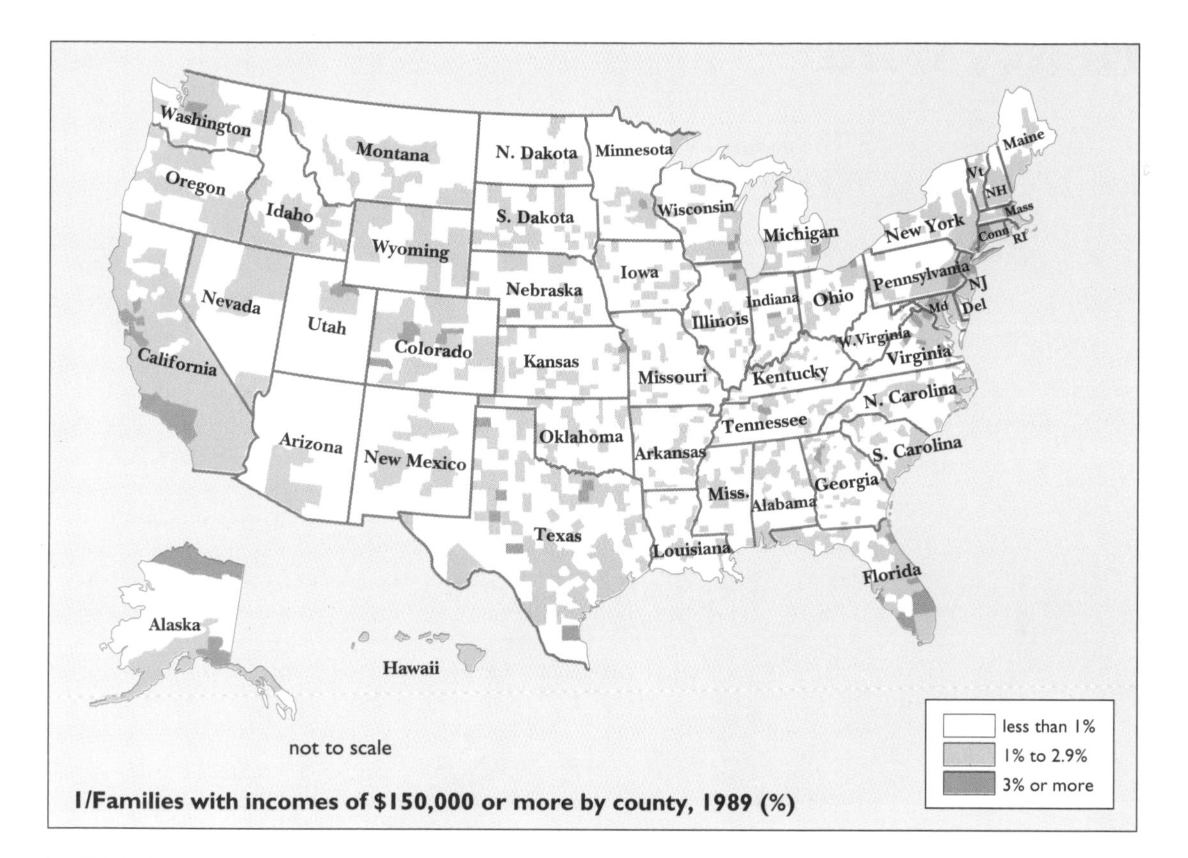

1/Families with incomes of $150,000 or more by county, 1989 (%)

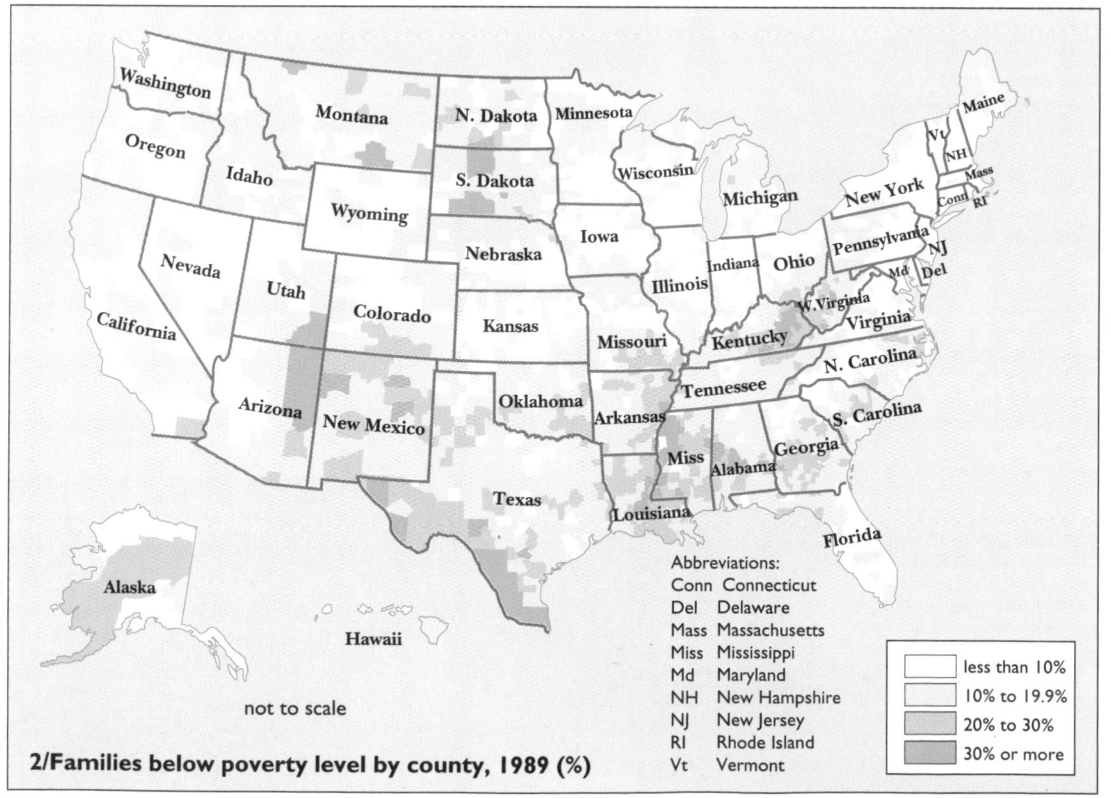

2/Families below poverty level by county, 1989 (%)

Henry Ford

The automobile industry revolutionized American life, and Henry Ford's Model T set the pace.

"The customer can have a Ford any color he wants—so long as it's black."
Henry Ford

Ford was a Michigan farmboy who hated farm work, and loved to tinker with machines. Despite having little formal education, and no use for books or high-falutin' ideas, he founded the Ford Motor Company in 1903. By 1920, every other car in the world was a Ford Model T.

His first successful automobile appeared in 1896. From the start, he had little interest in making cars for the luxury market. He scorned expensive fancy automobiles. "I will build a motor car for the great multitude". The first Model T, a utilitarian vehicle ideal for the unpaved roads of rural America, was introduced in 1909. It sold for $950, and had the virtue of being simple in design. The American male love affair with the automobile began with the joys of tinkering with the Model T.

The Duryea Brothers built the first gasoline-powered automobile in 1893. Henry Ford (right, in 1896)*, seen at the controls of an early automobile, began his career in the industry with $28,000 in capital. By the time the portrait of Ford* (above right) *was taken in the 1920s, he was the most famous manufacturer in the world. His success was built upon the Model T* (far right)*, which carried the Ford name on its radiator. The driver had an imposing horn to announce the car's presence (operated by a vigorous squeeze of the rubber ball), and was started by a crank handle. Automatic electric ignition came in the 1930s and was among the most welcome of all technological innovations for the automobile.The automobile was a significant factor in the sexual revolution of the 1920s. Preachers warned the impressionable young about the dangers of smoking, speakeasies and the back seat of the Model T.*

By 1914, Ford's skilled workmen were able to produce a Model T in 14 hours. After he introduced the first moving assembly line at his Highland Park plant in Michigan, an entire automobile was completed in 93 minutes. The cost of a Model T fell to $290 in 1924. Cars rolled off his assembly line every ten seconds. The Model A, a more comfortable vehicle available in a range of colours, was introduced in 1927. Annual automobile registrations in America rose from 6.7 million in 1919 to 23 million in 1929.

Ford thought his workers should have no need of unions. He created a "Sociological Department" at his plants to look after his employees, and to keep an eye on their behaviour. He paid high wages and demanded sober, obedient workers. He introduced the $5 day to American industry in 1914. Ford's rival, Walter Chrysler, built the elegant Chrysler Building in New York City in 1929 to celebrate the success of his motor company. Ford, who hated Wall Street, cherished rural American values. No one did more to disturb the slow pace and isolation of rural life than Henry Ford.

The Spread of American Culture

From being an importer of culture in the 19th century, Americans have become the world's greatest exporter of images, values and the cultural "product".

Throughout the 19th century *nouveaux riches* Americans went to Paris and Florence to purchase works of art. Their wives bought the season's *haute couture* in Paris. American scholars and scientists left Harvard and Yale to study at German universities. Marriageable American heiresses—Edith Wharton's "buccaneers"—went on shopping sprees at the country estates of the English aristocracy. European values, taste and attitudes towards social hierarchy were copied by the American élite. The streets in the vicinity of Central Park in New York City were lined with mansions in English and Italian Renaissance, Georgian, and French baroque styles. Social life at the most exalted levels became ever more aggressively European.

A century later American culture strides the planet. Indignant fundamentalists denounce wet T-shirt competitions and McDonald's; the satellite dishes which bring Bay Watch to the television screens of Teheran are ordered to be removed; French ministers splutter at the increasing Americanisms in a language prideful of its purity. American films and TV programmes appear throughout the world.

The "American Century" announced in 1941 by Henry Luce, publisher of *Time, Life* and *Fortune*, is our reality. For Luce the "American Century" was the age of General Motors and Standard Oil; for ourselves it is something less tangible than U.S. Marines guarding the imperilled assets of American corporations. It is the projection of an idea of America, of a style which has not yet lost its rich capacity to agitate custodians of traditional ways of life.

It was something a gentleman did. The collecting of works of art, as well as glass, carpets, stuffed animals, trophies etc., became something of a full-time occupation for the super-rich. William H. Vanderbilt, who had two of New York's most spectacular private residences built on Fifth Avenue in the 1880s, owned paintings by Turner, Alma-Tadema, Millais, Meissonier and Corot. The public—strictly by invitation—was allowed to view the Vanderbilt art collection (below) one day a week.

The spread of American 'culture' was usually a euphemism for the spread of American corporate marketing. What could be sold to Americans in Ohio could also be sold to street boys in Cambodia (Pepsi in hand) and to young mothers in Hong Kong's New Territories (Coca Cola with straw). American icons like the baseball hat and Levi's became the visual symbols of the young. But the heavily-promoted American football and baseball remained little more than a curiosity. There were evident limits to American cultural influence. McDonald's led the American commercial rush to introduce fast food into the very heart of the once forbidden Communist and post-Communist world. Beijing, (above left), and the Arbat in Moscow, (above right), were early targets.

The idea of "American Century" is contradicted by a triumphant economic position of the United States in the 1990s. The greatest creditor nation in 1945—the benefactor and bankroller of European reconstruction—expired during the reign of Ronald Reagan. American industrialists won corporate bonuses for exporting jobs to low-wage Third World countries; American capitalists bought junk bonds; and the communities which were at the heart of the productive colossus of the modern world have been devastated. There was little sympathy for 'Rust Belt' America in the 'Sun Belt' south and west. In the 1980s, one third of all industrial jobs disappeared from the United States economy. Toyotas, Nissans and a few BMWs fill the car parks of suburban shopping malls in Illinois. There are sweatshops once again in Los Angeles and New York. The sense of powerlessness, of being at the mercy of foreign and inimical interests (which lies at the heart of foreign anxieties at "Americanization"), is felt as strongly in the home of Burger King (owned by a British conglomerate) and Rockefeller Center (owned by a Japanese conglomerate) as in the rest of the free world.

Further Reading

NARRATIVE AND GEOGRAPHICAL HISTORIES

Brogan, H., *The Pelican History of the United States of America,* Penguin Books, 1986

Bumsted, J.M., *The Peoples of Canada,* Oxford University Press, 1992

Conrad, M., Finkel, A and Jaenen,C. *History of the Canadian Peoples,* vol. 1 *Beginnings to 1867*; vol. 2 *1867 to the Present,* Copp, Clark, Pitman, 1993

Meinig, D.W., *The Shaping of America,* vol. 1 *Atlantic America 1492–1800,* vol. 2 *Continental America 1800–1867,* Yale University Press, 1986, 1993

Meyer, M.C. and W.L. Sherman, *The Course of Mexican History,* Oxford University Press, 1991

Miller, R.R., *Mexico: A History,* University of Oklahoma Press, 1985

Mitchel, R.D. and P.A. Groves, eds., *North America: The Historical Geography of a Changing Continent,* Hutchinson, 1987

Morison, S.E., *The Oxford History of the American People,* Oxford University Press, 1965

ATLASES AND REFERENCE BOOKS

Asante, M.K. and M.T. Mattson, *Historical and Cultural Atlas of African Americans,* Macmillan, 1992

Beck, W.A. and Y.D. Haase, *Historical Atlas of the American West,* University of Oklahoma Press, 1989

Ferrell, R.H. and R. Natkiel, *Atlas of American History,* Facts On File, 1987

Homberger, Eric, *The Historical Atlas of New York City,* Henry Holt, 1994

Morris, R.B., *Encyclopedia of American History,* rev. ed., Harper & Row 1965

PART I TO 1492

Brotherston, G., *Images of the New World,* Thames & Hudson, 1979

Coe, M.D., *The Maya,* Thames & Hudson, 1966

Davies, N., *The Aztecs,* University of Oklahoma Press, 1980

Hagen. W.T., *American Indians,* University of Chicago Press, 1961

PART II TO 1775

Crosby, A.E., Jr., *The Columbian Exchange,* Greenwood, 1972

Eccles, W.J., *Essays on New France,* Oxford University Press, 1987.

Gibson, C., *Spain in America,* Harper & Row, 1967

Quinn, D.B., *North America from the Norse Voyages to 1612,* Harper & Row, 1977

Rawley, J.A., *The Transatlantic Slave Trade,* Norton, 1981

PART III TO 1861

Singletary, O., *The Mexican War,* University of Chicago Press, 1960

Stampp, K.M., *The Peculiar Institution: Slavery in the Ante-Bellum South,* Alfred A. Knopf, 1956

Wood, G.S., *The Creation of the American Republic, 1776–1787,* Norton, 1972

PART IV TO 1898

Stilgoe, J., *Metropolitan Corridor: Railroads and the American Scene,* Yale University Press, 1983

PART V TO 1945

Dulles, F.R., *America's Rise to World Power 1891–1954,* Harper & Row, 1955

Hofstadter, R., *The Age of Reform from Bryan to F.D.R.,* Alfred A. Knopf, 1955

Knight, A., *The Mexican Revolution,* 2 vols., Cambridge University Press, 1986

Wiebe, R.H., *The Search for Order 1877–1920,* Macmillan, 1967

PART VI TO PRESENT

Blum, J.M., *Years of Discord: American Politics and Society, 1961–1974,* Norton, 1991

Crockatt, R., *The Fifty Years War: The United States and the Soviet Union in World Politics, 1941–1991,* Routledge, 1995

Glazer, N. and Moynahan,D.P. *Beyond the Melting Pot,* Massachusetts Institution of Technology Press, 1963

Harrington, M., *The Other America,* Penguin Books.

Index

References shown in **bold** are maps or pictures. Quotes are in *italics.*

GENERAL

PLACES

Acknowledgements

Picture Credits

British Library: 14, 85t

British Museum: 16t

Chicago Historical Society: 87

Codex: 15bl, 15br

E.T. Archive, London: 29b (James Isham), 31tl, 45b, 62

Edith La Francis Collection (we have made every effort to obtain copyright for these but have been unsuccessful): 71tr, 71b

Eric Homberger: 28, 46, 49, 63tr, 70, 93l, 112, 127

Ferdinand Anton, Munich: 18b, 34, 36t

George Eastman House (International Museum of Photography): 65tr

New York Bound Bookshop, New York: 93r, 95, 97

Library of Congress: 106

Mary Evans Picture Library, London: 99tl, 99tr, 103tr

Michael Holford: 15t

New Haven Colony Historical Society: 71tl, 126

'*New York in the Nineteenth Century*', by John Grafton, (Dover Pictorial Archives series): 136

Peter Newark's American Pictures, Bath: 23, 29tl, 29tr, 30, 31tr, 32, 33, 40, 42t, 43, 47, 50, 51, 53, 54t, 58b, 60, 64, 65bl, 69, 80, 81, 83, 85b, 86, 90, 91, 96, 101, 102, 103tl, 105tl, 105tr, 105cr, 108, 111, 123, 131, 132

Popperfoto: 98, 110, 114–115, 116, 117, 118, 119, 121, 134, 135t

Professor Keith Roberts, Norwich: 92

Provincial Archives of Manitoba, Winnipeg, Canada: 35b

Robert Harding Picture Library, London: 22 (Robert Frerck), 25, 135b, 128 (FPG International), 129 (FPG International), 72 (FPG International), 73 (FPG International), 74 (FPG International), 75 (FPG International), 76 (FPG International), 77 (FPG International), 78 (FPG International), 79 (FPG International), 94 (FPG International), 124, 137br,

Smithsonian Institution, Washington, D.C., (Department of Anthropology): 27

Sotheby and Co.: 17

The Bridgeman Art Library, London: 24 Bibliotheque Nationale, Paris; 35t Private Collection; 36b Biblioteca Nacional, Madrid; 38 British Library, London; 39 British Museum, London; 48 British Library, London; 52 Pennsylvania Academy of Fine Arts, Philadelphia; 54b Cadogan Gallery, London; 55 Private Collection; 56 Chateau de Versailles, France; 57 Yale University Art Gallery, New Haven; 58t private collection; 63tl Museo del Virreinato, Mexico; 67 Victoria and Albert Museum, London; 100 Museum of the City of New York; 68 private collection; 109 Equitable Life Insurance Company, New York

The Fotomas Index: 41, 44, 45t

The Robert and Lisa Sainsbury Collection, University of East Anglia, Norwich. Standing Figure of a Ball Player. Central America AD 300–600. Photo: James Austin: 22

Tony Stone Images: 120, 137bl

University of Washington Libraries, Seattle: 84

US Geological Survey, Reston, Virginia (Department of the Interior): 88

Werner Forman Archive: 16bl (Peabody Museum, Harvard University, Cambridge, MA), 16br (Museum of the American Indian, Heye Foundation, New York), 18tl (Field Museum of Natural History, Chicago), 18tr (Field Museum of Natural History, Chicago), 19bl (private collection, New York), 19br (private collection, New York)

For Swanston Publishing Limited

Concept:
Malcolm Swanston

Editorial:
Caroline Lucas
Chris Schüler

Editorial Assistance:
Rhonda Carrier
Stephen Haddelsey

Illustration:
Ralph Orme

Cartography:
Andrea Fairbrass
Peter Gamble
Elsa Gibert
Elizabeth Hudson
Isabelle Lewis
David McCutcheon
Kevin Panton
Peter Smith
Andrew Stevenson
Malcolm Swanston

Additional Cartography:
Advanced Illustration, Cheshire

Index:
Jean Cox
Barry Haslam

Typesetting:
Jeanne Radford

Picture Research:
Eric Homberger
Caroline Lucas
Charlotte Taylor

Production:
Barry Haslam

Separations:
Central Systems, Nottingham.
Quay Graphics, Nottingham.